Logistics Systems Analysis

4th Edition

Carlos F. Daganzo

Logistics Systems Analysis

Fourth Edition
with 70 Figures
and 4 Tables

 Springer

Professor Carlos F. Daganzo
Institute of Transportation Studies
416 McLaughlin Hall
University of California
Berkeley, CA 94720
USA
E-mail: daganzo@ce.berkeley.edu

Cataloging-in-Publication Data
Library of Congress Control Number: 2005924096

ISBN 3-540-23914-6   Springer Berlin Heidelberg New York

Springer is a part of Springer Science+Business Media
springeronline.com

© Springer-Verlag Berlin Heidelberg 2005
Printed in Germany

Cover design: Erich Kirchner
Production: Helmut Petri
Printing: Strauss Offsetdruck

SPIN 11356912   Printed on acid-free paper – 42/3153 – 5 4 3 2 1 0 to Val

Preface to the Fourth Edition

This expanded edition of "Logistics Systems Analysis" includes new research results and numerous modifications to enhance comprehensiveness and clarity. It has two new sections, a new appendix, and more than half a dozen new figures. A few references have also been added, but the bibliography is not exhaustive. Much of the new material is based on work by Profs. Alan Erera (Georgia Tech), Karen Smilowitz (Northwestern U.), and by PhD candidate Yanfeng Ouyang (U.C. Berkeley). Their help is gratefully acknowledged. The financial support of the National Science Foundation and the Volvo Foundations Center of Excellence for the Future of Urban Transportation at U.C. Berkeley is also acknowledged.

The new appendix presents the logic behind the traveling salesman and vehicle routing results used in Sec. 4.2 to describe the transportation operation; Chapter 4 is more self-contained as a result. New section 5.6 introduces and evaluates a general method that automatically translates the continuum approximation recipes of Chapters 4 and 5 into discrete system designs. This closes a gap in previous editions. Other additions include an explanation of how to develop system designs that can efficiently accommodate real-time control strategies to manage uncertainty (new section 4.6.3), and extensions of the many-to-many design ideas of Chap. 6 (in expanded section 6.5.3). An errata corrigendum will be posted on the authors's web site: http://www.ce.berkeley.edu/~daganzo/ This web site also explains how to order the solution manual to the problems in the book (professors only).

Carlos F. Daganzo
Berkeley, California
November 2004

Preface to the Third Edition

Aside from the removal of minor errors, the main modification in this printing of "Logistics Systems Analysis" is an improved explanation of many passages that had been found confusing by colleagues and students, most notably in chapters 5 and 6. A few references have also been added, mostly having to do with material complementary to that of the book.

I would like to acknowledge the comments of Profs. Eric Mohr, Wei Lin, and David Lovell, and the input of graduate students Flavio Baita, Alan Erera, Reinaldo Garcia, and Juan Carlos Muñoz.

Carlos F. Daganzo
Berkeley, California
November 1998

Preface to the Second Edition

The presentation and ideas in this second edition of "Logistics Systems Analysis" remain essentially unchanged from those of the first edition. The main modifications are the inclusion of an index, a more current reference list with brief discussions where appropriate, editorial changes to improve clarity, and the removal of a number of errors that had crept in the 1991 edition.

After teaching from this monograph for a number of years, I have found that students do not really master the material in it until they have used it to *formulate* and solve real life problems of interest to them. Although solving a number pre-formulated problems is no substitute for this experience, such an effort can be a useful first step toward the ultimate goal. In view of this, an informal set of solutions to some of the exercises listed at the end of each chapter has been developed. They can be ordered by writing to the Institute of Transportation Studies, Publications Office, 109 McLaughlin Hall, University of California, Berkeley, California, 94720.

My sincere gratitude goes to Mrs. Ping Hale for her skillfull preparation of the revised manuscript, and to the University of California Transportation Center for funding our efforts.

Carlos F. Daganzo
Berkeley, California
August, 1995

Preface to the First Edition

Logistics, the subject of this monograph, is narrowly defined here to be the science that studies how to convey items from production to consumption in cost-effective ways; some subjects of interest to logistics managers such as reliability and maintenance are not addressed. The theories that are covered, on the other hand, apply to generic items that can represent people, as well as freight; they should be of interest to passenger transportation firms and agencies.

Besides transportation, a logistics system usually includes other activities such as inventory control, handling, and sorting, which must be carefully coordinated if cost-effectiveness is to be achieved. Yet, both in theory and practice these activities are often examined separately.

The operations research field includes sub-fields with specialized journals in inventory control, transportation, warehousing, etc... Over the years, these sub-fields have evolved into disciplines that have developed their own specialized conventions and jargon, as a result making it increasingly difficult for researchers to communicate across disciplinary boundaries. Something similar happens in practice when firms become compartmentalized; if responsibilities for different logistical activities are allocated to different managers, decisions in the best interests of the firm are difficult (if not impossible) to make.

This monograph represents an attempt to examine logistics systems in an integrated way. By necessity, we will not represent any of the activities as precisely as would be done in each one of the sub-fields of *OR,* but we will try to model them accurately enough to capture their essence. Our goal is to describe, and show how to find, rational structures for logistics systems, including their operation and organization.

This monograph also departs from traditional operations research procedures in that it tries to avoid detailed descriptions of both problems and their solutions. For a typical problem, instead of searching for the ultimate solution based on reams of detailed data and time consuming numerical analyses, our goal will be to present reasonable solutions (described in terms of their properties) with as little information as possible. In fact a goal of our analyses will always be to determine what is the *least* amount of information that is needed to make a rational decision, and to use the simplest most transparent approach possible to identify good solutions. These features of our approach can help overcome the decision-makers' natural distrust of "black-boxes", and be quite helpful in instances where time is of the essence. This is not to say that the more traditional detailed approaches to problem solving should not be used; when time and information availability allow it, numerical detailed methods have proven to be quite useful. Yet, even in these instances detailed solutions sometimes can be improved if they are preceded by an exploratory analysis as described in this monograph.

The work presented in this monograph is a result of efforts undertaken by this author, his close colleagues and students for the last decade.

In the early eighties, Dr. Larry Burns of the General Motors Research Laboratories and I became interested in the internal movement of goods of a large firm. Because transportation in a large firm shares many similarities with public transportation, we realized that, as had been done in the 70's for transit design problems and location problems by Prof. Newell (of the University of California, Berkeley) and his students, it was possible to substitute large numbers of data by suitable averages and to treat discrete problems in a continuous manner. We did not suspect at the time the impact that this endeavor was to have on General Motors. (To date, and using this approach, the General Motors research team headed by Dr. Burns has been commended repeatedly for numerous logistics improvements at *GM*).

In an effort to formalize this thinking, Newell and I taught an advanced 1 unit graduate seminar at U.C. Berkeley in 1982. Later that year I expanded and tested these notes while on sabbatical leave at M.I.T., hosted by Prof. Yosef Sheffi. Since then work has continued, with the contribution of Prof. Randolph Hall of U.C. Berkeley being particularly noteworthy, and the most current ideas are now taught in a four unit graduate course on networks and logistics.

Although most of the ideas in the course have been documented in the open literature, Professor Martin Beckmann (of Brown University) remarked at an *EURO/TIMS* conference in Paris in the late 80's that it is not easy for an "outsider" to get an overall view of this work. He convinced me that too many journals have published articles on the subject (sometimes with an unnatural chronology), and thus planted the seed for this monograph in my mind.

Mostly based on published works, this monograph attempts to present the subject in a logical way. New ideas are also presented when, in order to tell a cohesive story, "voids" in the literature had to be filled. Voids still remain and, hopefully, the monograph will spur further work on this young and evolving subject.

The first two chapters introduce preliminary ideas. Chapter 1 illustrates, by means of an example, the problem solving philosophy of the monograph, and Chapter 2 explains the accounting method for logistics costs. Chapters 3 through 6 describe the theory as is applied to gradually more complex problems. Chapter 3 explains the optimization method in detail and illustrates it with a problem involving only one origin and one destination. Chapter 4 examines problems with one origin and many destinations (or vice versa), assuming that each item travels in only one vehicle; Chapter 5 allows for transshipments at intermediate terminals. Finally, Chapter 6 examines "many-to-many" problems.

Before getting started, some remarks about notation and organization need to be made. Equations, tables and figures will be numbered *(a.b)*, where *"a"* is the chapter number and *"b"* the equation number. Also, because an attempt has been made to use as consistent a set of symbols throughout the monograph as possible, the reader should expect the notation often to be quite different from that in the references, which is unavoidable. Each chapter begins with some brief remarks on a few recommended readings that are closely related to the topic at hand, and ends with a set of suggested exercises and a list of symbols. A reference list with the bibliographic citations is provided at the end of the monograph.

Although not its main goal, the monograph could be used for teaching graduate students about logistics, perhaps in a course also covering the more traditional *OR* optimization tools. It should be possible to cover the most basic ideas in 10 one hour classes with about two classes per chapter, but a lengthier exposition is recommended to delve into details. To this effect, each chapter contains a few suggested exercises, intended to solidify the students' grasp of the concepts in the chapter and/or explore extensions that could not be discussed in the text.

In closing, I would like to thank my friends and colleagues, Profs. Gordon Newell and Randolph Hall of the University of California (Berkeley), and Dr. Lawrence Burns of General Motors Corporation, for their contribution to the ideas in this monograph.

I also thank Mrs. Phyllis DeFabio and Mrs. Ping Hale for their patience and perseverance in preparing the various versions of this manuscript. Ms. Gail Feazell prepared most of the Figures. The support of the Institute of Transportation Studies is also gratefully acknowledged. But most of all, I appreciate the love and understanding of the beauties and beauticians of my life.

Table of Contents

1 The Use of Succinct Models and Data Summaries

Readings for Chapter 1

As we do in this chapter, Blumenfeld et al. (1987) describe the advantages of simple models; the opinions expressed in this reference are based on a case study where succinct models based on data summaries proved very effective; the reference is easy to read. Newell (1973) argues that a family of related transportation and location problems can be solved approximately with an approach that ignores "details"; this paper was the "seed" for the continuum approximation method to be presented in Chapter 3.

1.1 Different Approaches for Solving Logistics Problems

Logistics can be defined in a number of ways, depending on one's view of the world; but in this monograph it is taken to be the set of activities whose objective is to move items between origins and destinations (usually from production to consumption) in a timely fashion.

Traditionally logistics problems are solved by gathering as much detailed information as possible about the problem, formulating a mathematical program including as input data all the information that *might* possibly be relevant, identifying solutions in detail by means of numerous decision variables, and then using the computer to sort through this numerical maze. Because the data collection effort can be onerous, sometimes decisions are made with no systematic analysis. On other occasions the numerical optimization problem is *NP-hard* (which makes it difficult to obtain good solutions when the problem is formulated without simplification) and decisions are made on the basis of heuristic solutions, which are not particularly insightful.

This monograph presents an alternative approach for logistics systems planning and analysis where, even if detailed data are available, the bulk of the approximations are made right at the outset. Detailed data are replaced by concise summaries, and numerical methods are replaced by analytic models. Without numerous pieces of information the analytic models can be solved accurately; it becomes possible to identify broad properties of solutions close to the global optimum. These near-optimal solutions are then used to formulate guidelines for the design of implementable solutions; i.e. solutions that satisfy all the detailed requirements ignored in the analysis. This final "fine-tuning" step may be carried out with the help of a (more traditional) computer optimization program, but this is not a requirement. The remainder of this chapter compares the traditional and proposed approaches. Daskin (1985) reviews both schools of thought, although the non-detailed approach has evolved a great deal since then. A more recent review of the latter is given in Langevin et.al (1995).

The proposed approach is particularly useful in planning applications over long horizons, where data are uncertain. Its simplicity also allows analysts to develop qualitative insights (e.g. into the most important trade-offs influencing the choice of final solutions) which can then be communicated meaningfully, without interfering details, to managers and decision-makers. Better decisions will be made as a result. A clear understanding of the trade-offs at work is also important because it allows quick but educated decisions to be made on the spot despite a changing world. Should conditions such as labor prices, availability of locations, etc... change suddenly between design and implementation, alternative solutions that account for these changes readily become apparent because the basic design principles stay the same. This is fortunate because sometimes it is just not possible to prepare new data and interpret the result of another numerical optimization in a timely manner.

The following example, reminiscent of the case study in Blumenfeld et al. (1987) and representative of the kinds of problems addressed later in this monograph, is used to expand the scope of the comparisons between detailed and simple models. It focuses on accuracy issues and illustrates how broad decisions can be evaluated without detailed data.

1.2   An Example

A hypothetical manufacturer of computers, radios and television sets has three factories and 100 distribution centers in the continental U.S. The factories are located in Green Bay (computer modules), Indianapolis (televisions, monitors and keyboards) and Denver (consoles); see Figure 1.1. Some of these components must be assembled before they are sold, and this can be done either at the distribution centers themselves or at a central location next to the Indianapolis factory, which we shall call "the warehouse". We seek a distribution strategy that will minimize the sum of the transportation and inventory costs per year; see Figure 1.2.

The computer modules cost $ 300 and weigh 5 lbs, the televisions (as do the monitor/keyboard sets) cost $ 400 and weigh 10 lbs, and the consoles cost $ 100 and weigh 30 lbs. Road trucks can carry 30,000 lbs and can be hired with driver for $ 1 per mile. (In practice real costs are higher, and volume–not weight– often limits shipment size). We also assume that an inventory penalty is paid indirectly for each day that a product is waiting to be either transported or consumed. The penalty is figured to be 0.06% of the cost of the product per working day waited, corresponding to an interest rate of about 15% per 250–work–day year. Chapter 2 will discuss transportation, inventory, and other logistics costs in more detail.

Fig. 1.1   Location of 3 hypothetical factories

For a given distribution strategy, cost can be calculated if we know the locations of the distribution centers and the yearly demand at each destination by item type. This is the detailed information that was mentioned in Sec. 1.1.

Table 1.1 (at the end of the chapter) contains 300 entries giving the hypothetical distance in miles from each factory to each destination center. The locations of these centers were generated by random independent draws from a uniform distribution of positions in a 2500x1000 mile rectangle with sides parallel to the coordinate axes. Distances were calculated as the sum of the absolute differences in the points' coordinates.

For purposes of illustration we will assume that each distribution center sells every working day 10 television/console sets and 10 computer units consisting of a module, monitor and keyboard. (Although in practice these figures would change across distribution centers and from day to day, such complications are not included in our example because the distance table contains enough details to illustrate our point. See exercise 1.2.)

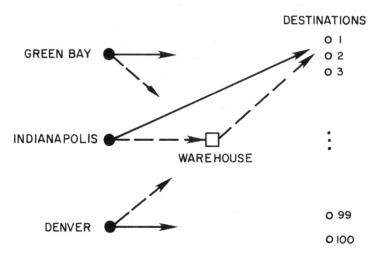

Fig. 1.2   Possible strategies for using a warehouse

We also know from observation of the map (or a cursory perusal of Table 1.1) that the average distance from a factory to a distribution center, as well as between the Green Bay and Denver factories and the warehouse, is on the order of 10^3. More accurate estimates could be obtained from the raw data, but this is not necessary.

Two broad strategies are considered first. With strategy (i) the warehouse is skipped, and everything is shipped directly from the factories to the centers in full trucks without intermediate stops. With strategy (ii) everything is assembled at the warehouse. Trucks still travel non-stop and full from the factories to the warehouse, and from the warehouse to the centers.

The total transportation cost per year is obtained as the product of the cost of a typical truck trip, $\$ 10^3$, and the number of truck trips per year. Every year, each destination requires 2500 items from the Denver and Green Bay factories and 5000 from Indianapolis. This is equivalent to $2500(30/30,000)=2.5$ trips per year from Denver, 0.417 from Green Bay and 1.667 from Indianapolis. And the total transportation cost is: $100(2500+417+1667) \approx 4.6 \times 10^5$ \$/yr.

This strategy minimizes transportation cost because every item travels the least distance possible in a full truck, but it exhibits exorbitant inventory costs. Because a truck from Green Bay only visits a center once every 0.417^{-1} years, items from Green Bay spend on average $0.417^{-1}/2$ years waiting to be consumed at the destination and a similar amount of time at the origin waiting to be loaded. This result assumes that Green Bay makes up the loads for all the destinations simultaneously. The experienced reader will recognize that the inventory at the factory could be virtually eliminated if loads were to be made for the different destinations in succession, e.g., as explained in Sec. 4.3.3 of the text, but the calculations that follow are simpler by assuming simultaneous load make-up for all the factories. This simplification is adopted because it does not change the qualitative conclusions significantly. Note as well that the long waits arising from the current strategy are impractical, and this is confirmed by the high inventory costs that result. Given the 0.417^{-1} year wait, the inventory cost per item delivered from Green Bay is: $(300)(.15)(0.417^{-1}) = \$\ 108$, and from the Denver and Indianapolis factories: \$ 6 and \$ 36.[1] The yearly inventory cost is thus a staggering: $(2500)(100)[108+6+2(36)] = 46.5 \times 10^6$ \$/yr, which brings the total to about 47 million dollars per year.

Strategy (ii) increases transportation costs because warehouse consolidation requires items to travel more miles. However, it can also decrease inventory costs. Because all the distribution centers are served from the warehouse, with its higher volumes shipped, more frequent delivery is possible. Since all trip distances are on the order of 10^3 miles, the total transportation cost between the warehouse and the centers should be the same as the total transportation cost for strategy (i): 4.6×10^5 \$/yr. Similarly, between the factories and the warehouse, the cost should be the same as for strategy (i), excluding the Indianapolis items: about $100(2500+417) \approx 3 \times 10^5$ \$/yr. The total is thus: 7.6×10^5 \$/yr.

Three components of the inventory cost should be considered; they are the cost of waiting between: (1) Green Bay and Indianapolis, (2) Denver and Indianapolis, and (3) Indianapolis (warehouse) and the destinations. (It is assumed that the item transfer from the Indianapolis factory to the warehouse does not result in an inventory cost). Green Bay ships 1.25×10^6 lbs/yr, which results in 41 truck trips per year to the warehouse, yielding an inventory cost of \$ 1.1 per item: $(300)(.15)(41^{-1}) \approx 1.1$. From Denver, trips to the warehouse are made 6 times more frequently since consoles are six times heavier than modules, and since in addition consoles are 3 times cheaper, the inventory cost per item must be 18 times smaller; i.e. \$ .06.

---

[1] We are assuming that the Indianapolis' shipments contain both TV's and monitor/keyboard sets.

On a yearly basis the inventory cost for freight inbound into the warehouse is thus: $(1.1 + .06)(2500)(100) \approx \$ 2.9 \times 10^5$.

Each center receives $(2500)(15+40)$ lbs worth of goods every year and as a result is visited by a truck 4.6 times a year. Insofar as loads are being made-up at the warehouse for all the centers simultaneously, the inventory cost per computer set delivered is $(300+400)(.15)(4.6^{-1}) \approx \$ 23$. For television sets it is \$ 16. Since 250,000 sets of each type are processed every year, the inventory cost for outbound freight from the warehouse is: $(23+16)(250,000) \approx 98 \times 10^5$ \$/yr.

The total inventory cost for strategy (ii) is 10.1×10^6 \$/yr. Compared with strategy (i), the large reduction more than offsets the small increase in the transportation cost. The total cost is reduced by a factor of 4, to about 10.9 millions of dollars per year. Given the outcome of strategy (i), it is natural to seek reductions in the inventory cost by shipping more frequently in partially full trucks. No doubt this will increase the transportation cost, but it should be intuitive that if the dispatch frequency is judiciously chosen the total cost will decrease. Chapters 2 and 3 explain how an optimal shipment frequency and the associated cost can be easily obtained for each origin destination pair. Here we only need the optimal cost formula as it applies to our example. It is a function of only three variables, the freight rate for one trip (in \$), the origin to destination flow (in lbs/yr) and the freight value (in \$/lb):

$$\text{cost/yr} = 2[(.15)(\text{rate})(\text{flow})(\text{value})]^{1/2}. \qquad (1.1)$$

Recall that the freight rate equals the distance (in miles).

With this knowledge, we can explore extensions to strategies (i) and (ii), labeled (iii) and (iv), in which an optimal shipment frequency is used on all the transportation routes. Note that implementation of these strategies requires careful coordination between the transportation and inventory control managers, if they are different individuals.

Using Eq.(1.1), the reader can verify with very little effort that the total yearly cost for strategy (iii), still using 10^3 miles per truck trip as a coarse estimate for all the distances, is about 6.8 million dollars. The cost estimate for strategy (iv) is about \$ 4.6 million.

The cost could conceivably be reduced if one allows some of the factories to ship direct and the remaining through the warehouse (strategy (v)); the calculations are simple since only two new routing schemes are possible: either Denver direct and Green Bay through the warehouse, or vice versa. The former is preferred but its cost, $\$ 5.5 \times 10^6$, is still higher than for strategy (iv).

In later chapters we will show how routing strategies in which trucks are allowed to make multiple stops can be analyzed in a simple manner with data summaries; for our particular problem they can reduce cost somewhat further.

The reader is challenged to verify that the costs obtained using the detailed data of Table 1.1 to evaluate costs and dispatching frequencies (in millions of dollars per year) are: 47 for strategy (i), 10.8 for strategy (ii), 6.7 for strategy (iii), 4.5 for strategy (iv), and 5.4 for strategy (v). Although these figures differ slightly from our rough estimates, the ranking of the strategies— (i) < (ii) < (iii) < (v) < (iv)— is preserved. Table 1.2 summarizes the total costs. The two cost columns in the Table are so close because the inventory cost, a large component of total cost for strategies (i) and (ii), is given exactly by the summary data. This would happen even if the consumption rates were to change across destinations. With different consumption rates there would be larger discrepancies for the remaining strategies, but their relative ranking would tend to stay the same; see exercise 1.2. This ranking is also robust to (moderate) errors in summary data. If, for example, the average trip distance had been overestimated by 25%, the yearly cost estimate for strategies (i) to (v) would have been (in millions of dollars): 47, 11, 7.6, 5.0 and 6.0. Clearly, decisions taken with the simple approach are cost-effective, even if the yearly costs are not precisely known.

Table 1.2   Summary of the results for the case example

Shipping Strategy	Estimated total cost in 10^6 \$/year	Exact total cost in 10^6 \$/year
(i) Direct, full trucks	47	47
(ii) Warehouse, full trucks	10.9	10.8
(iii) Direct, optimal frequency	6.8	6.7
(iv) Warehouse, optimal frequency	4.6	4.5
(v) Part direct, part warehouse	5.5	5.4

Note that in verifying the ranking each one of the 100 destinations must be considered separately. This is tedious, much like the data preparation effort needed for "real life" detailed numerical analyses. (To analyze a strategy allowing multiple truck stops one would have to deal with 103^2 distances – in order to account for the distances between all possible customer pairs.)

The reader may complain that we have not explored more complicated strategies. For example, strategy (v) can be generalized to allow some centers to be served through the warehouse and others directly, so that the locational advantages of specific destinations can be exploited. Granted, this additional flexibility has the potential for small improvements, but the price is the additional complexity of implementation. In any case, as explained in later chapters, even strategies that seem to require detailed information can often be examined with properly summarized data.

1.3   Remarks

This section discusses the merits of the proposed approach, using the experience gained from this example as an illustration. The approach is evaluated on two dimensions: the flexibility it affords, and the accuracy of its results.

1.3.1 Usefulness and Flexibility

We have already said that simple models reveal quite clearly the trade-offs underlying their recommendations; this is especially true when the final solution can be expressed in closed form as a function of only a few data summaries. In practice, a physical understanding of the reasons for the recommendations is quite useful to have because:

(1) it can be a convincing communication bridge between the analyst and the decision-maker when recommendations for action are being made, and

(2) it may point to better solutions than those allowed by the original formulation. This may lead to a reconsideration of the original question, perhaps suggesting that the scope of the problem should be expanded.

Although we did not write any analytical expressions in Sec.1.2. (this will be done in later chapters) the reasons for the appeal of strategy (iv) are clear. If no transshipments are made, and given the small flows between individual origins and destinations, then either inventory costs have to be large, as with strategy (i), or else vehicles have to travel nearly empty, as with strategy (iii). Transshipments at the warehouse eliminate nearly 1/3 of the vehicle routes and increase route flows, reducing the cost per item on each route dramatically. Strategies (ii) and (iv) exhibit lower total costs as a result, even though individual items travel double the distance. Because this seems like a steep travel penalty to pay for perhaps too much flow consolidation, we are entitled to suspect that less consolidation with a smaller distance penalty could be more economical. What if flows could be increased less dramatically, by a factor of 2 (say) instead of 3, without increasing the travel distances noticeably? Consideration shows that this can be achieved by using three warehouses (one near each factory) and allowing different centers to receive items through the most conveniently located warehouse. This realization indicates that the original problem formulation should be reconsidered.

We had also mentioned earlier that insights gained during the analysis, especially if it yields a simple formula, reveal at a glance how the optimal solution should change if the basic conditions change. This knowledge should be most useful in a decision-making environment where conditions may change on short notice. In our case, the inefficiencies from low flows are most noticeable for the Green Bay factory, which makes expensive light items, and least so for the Denver factory. Thus, had the demand of television sets been much greater (which increases the flows from Denver and Indianapolis but not from Green Bay), we would expect strategy (v) to be the winner. The small and expensive items from Green Bay could be conveniently consolidated with the already heavy flow out of Indianapolis, and the bulky Denver items could be shipped directly. It should not be difficult to see (without any need for complex additional calculations) how the best distribution pattern should be modified if other changes occur, involving for example: inventory rates, transportation rates, item values, and/or demand rates.

1.3.2 Accuracy of Its Results

Before turning our attention to accuracy issues it is worth noting that the largest cost reductions in the example [e.g. from strategy (i) to (ii)] result from the simplest decisions [e.g. "consolidate everything at the warehouse"] for which little information and analysis effort is needed. Further improvements become gradually harder and harder to achieve both in terms of information, computation and difficulty of implementation. This pattern, typical of many applications, underscores two related facts: (a) lack of accurate detailed data should not be an excuse for avoiding analysis, and (b) precision should not be pursued at all costs for its own sake. With this as a prelude, we now discuss the accuracy of the traditional and proposed approaches.

Errors to the proposed approach arise primarily from the simplifications made to the data, while errors to the traditional approach arise mostly from failure of the algorithms to identify good solutions. In a way, thus, we have to choose between solving a model based on approximate data accurately, or a more precise model approximately. In the former case the approximations are explicit and can be easily interpreted; *non-essential information is eliminated early, before it can cloud the issues at hand.*

Another source of errors, ignored in the above remarks, arises from errors in the data (detailed or summarized) as they propagate to the final solution. While errors in data summaries can be easily traced to the final solution, the same cannot be said for models based on numerous detailed data. In later chapters we will discuss the perhaps surprising effect that errors in the data have on both detailed and simple models.

Besides data related errors, we must also consider the impact of problem formulation assumptions on both approaches; for example we may have to define how transportation costs depend on flow, how inventory costs depend on time, etc... If approximations are made in a detailed model the analyst has no simple way of assessing their implications. Thus, being aware that a certain assumption allows more powerful algorithms to be used, the analyst may be tempted to adopt it perhaps thinking: "what I lose in accuracy, I will gain in increased closeness to optimality". This is a great danger for, on occasion, seemingly innocent changes radically change the nature of the optimal solution. [For our example, if the transportation cost per trip is assumed to be proportional to the amount of freight carried, instead of a fixed quantity, then the optimal solution is to ship every item direct from the origin to its destination as soon as it has been produced. There is no incentive whatsoever to consolidate loads either temporally or spatially.]

The above discussion is not meant to suggest that detailed models should not be used. Rather, it should underscore that both approaches to logistics systems analysis have proper application contexts.

In an environment where changing political pressures (which are hard to quantify) and changing economic opportunities interfere with well planned *strategic* decisions, numerical solutions may be too rigid. A set of simple guidelines, coupled with some understanding of the relationship between key input factors and the type of solution that is desirable, may be more useful than a detailed analysis. Simple models are also recommended for planning applications; especially in the early stages of analysis (when the problem itself is not yet fully understood) because many different options and formulations can be explored with them. Detailed models, on the other hand, are most useful for finalizing decisions. After a strategic decision has been made, hopefully based on sound technical advice, the design guidelines obtained with simple models must be translated into a practicable solution that incorporates all the details. It is for this last step (termed "fine-tuning" in this monograph) that detailed computer optimization tools seem be most useful. Although this monograph will emphasize model building and guideline development, it will briefly describe some systematic fine-tuning methods as well.

Instead of being technique oriented (as most *OR* books are), this mono-graph is organized along problem lines and uses a single problem solving philosophy. Increasingly difficult problems are considered in each suc-ceeding chapter, using an incremental approach for their treatment – the insights and formulas developed in every chapter become "building blocks" for the ensuing ones. Chapter 2 discusses the various logistics costs considered in this monograph, and in the process also introduces the simple lot size model.

Suggested Exercises

1.1   Provide a 1 page description of each one of the recommended read-ings. Comment, in particular, on the problem solving philosophy of those two works as it relates to this chapter. Discuss points of agreement and disagreement as you see them.

1.2   Generate a table similar to Table 1.1 with a computer spreadsheet but now also including a different demand rate for each destination. Arrange the information with one row per destination. To generate the demands, as well as two sets of X- and Y-coordinates, use suit-able expressions involving the random number generator of your spreadsheet, and store each set on a separate column. Given the co-ordinates of three factories (stored elsewhere on the spreadsheet to-gether with the rest of the information on prices and weights, etc...), fill in the columns with the distances. Then, repeat the analysis of Sec. 1.2. using the average demand and the average distances (or any coarse approximation thereof) from the table you generated. You will find that, with variable demand, the numerical results of the detailed and simple approaches differ somewhat more signifi-cantly than in Sec. 1.2. Why? Does that change the ranking of the strategies? (For a uniform distribution of demand rates, the differ-ences between the detailed and simple approaches with the data in Table 1.1 remain below 8%).

Table 1.1   Production characteristics of the three Factories, and distances to all other points

Production Characteristics For the Three Factories			
	Green Bay	Indianapolis	Denver
price ($)=	300.00	400.00	100.00
weight (lbs)=	5.00	10.00	30.00
units/yr=	250000.00	500000.00	250000.00
Distance (Miles) Between Factories			
	Green Bay	Indianapolis	Denver
Green Bay	0.00	400.00	1100.00
Indianapolis	400.00	0.00	1100.00
Denver	1100.00	1100.00	0.00
Distance (Miles) To the 100 Distribution Centers			
Center Number	Green Bay	Indianapolis	Denver
1.00	584.71	648.56	515.29
2.00	1409.37	1409.37	309.37
3.00	2119.45	2119.45	1019.45
4.00	341.21	354.30	1441.21
5.00	861.41	1261.41	473.65
6.00	1363.68	1363.68	444.09
7.00	857.30	570.52	1357.30
8.00	937.19	537.19	1437.19
9.00	1222.73	1588.18	688.18
10.00	1875.46	1875.46	775.46
11.00	1385.30	1785.30	885.30
12.00	2070.18	2070.18	970.18
13.00	860.75	860.75	139.25
14.00	2214.84	2214.84	1114.84
15.00	526.86	526.86	966.51
16.00	1707.24	1707.24	607.24
17.00	152.37	247.63	958.95
18.00	1294.24	1659.63	759.63
19.00	641.85	641.85	774.15
20.00	845.44	845.44	952.46
21.00	1655.95	2055.95	1155.95
22.00	1513.14	1513.14	413.14
23.00	399.68	350.98	1450.98
24.00	1348.81	1348.81	655.88
25.00	407.64	253.78	1353.78
26.00	1001.02	1401.02	501.02

Table 1.1   (continued)

	Distance (Miles) To the 100 Distribution Centers		
Center Number	Green Bay	Indianapolis	Denver
27.00	997.20	997.20	755.62
28.00	443.71	58.19	1041.81
29.00	168.69	231.31	1148.58
30.00	501.99	501.99	853.83
31.00	1010.97	610.97	1510.97
32.00	571.78	971.78	542.24
33.00	795.88	841.63	304.12
34.00	2171.06	2171.06	1071.06
35.00	546.46	946.46	743.50
36.00	697.64	660.46	1197.64
37.00	231.72	231.72	1331.72
38.00	1363.63	963.63	1863.63
39.00	1964.34	1964.34	864.34
40.00	112.04	307.45	987.96
41.00	820.00	820.00	1920.00
42.00	1146.20	1146.20	198.53
43.00	1963.72	1963.72	836.72
44.00	888.51	488.51	1586.05
45.00	362.96	326.04	1426.04
46.00	1627.74	2027.74	1127.74
47.00	911.78	911.78	1048.26
48.00	1615.09	2015.09	1115.09
49.00	1262.57	862.57	1762.57
50.00	2170.74	2170.74	1070.74
51.00	924.65	524.65	1424.65
52.00	99.92	499.92	1178.72
53.00	710.23	1110.23	628.25
54.00	1677.30	2077.30	1177.30
55.00	827.57	427.57	1327.57
56.00	1327.32	1727.32	827.32
57.00	647.77	1047.77	739.44
58.00	320.38	418.26	779.62
59.00	1641.10	1783.92	883.92
60.00	657.25	657.25	669.32
61.00	800.39	800.39	1202.51
62.00	968.10	1290.45	390.45
63.00	2377.68	2377.68	1277.68
64.00	1443.41	1601.62	701.62
65.00	1402.79	1802.79	902.79
66.00	1853.90	1853.90	753.90

Table 1.1   (continued)

	Distance (Miles) To the 100 Distribution Centers		
Center Number	Green Bay	Indianapolis	Denver
67.00	1556.15	1852.01	952.01
68.00	2000.05	2000.05	900.05
69.00	915.33	915.33	612.42
70.00	1235.29	1235.29	422.84
71.00	1116.23	1116.23	664.63
72.00	923.32	523.32	1466.74
73.00	1241.57	1284.57	384.57
74.00	1036.80	636.80	1536.80
75.00	836.46	836.46	306.70
76.00	942.68	1342.68	442.68
77.00	107.41	507.41	1130.05
78.00	1295.62	1295.62	591.75
79.00	1952.37	1952.37	852.37
80.00	1312.65	1528.55	628.55
81.00	966.23	966.46	133.77
82.00	710.83	720.83	999.92
83.00	256.57	301.83	1356.57
84.00	1090.11	1490.11	590.11
85.00	703.16	703.16	1803.16
86.00	928.37	528.68	1428.37
87.00	588.68	588.68	991.56
88.00	1280.04	1280.04	473.06
89.00	1893.91	1893.91	793.91
90.00	471.56	871.56	851.39
91.00	508.87	508.87	1608.87
92.00	340.93	340.93	1440.93
93.00	427.53	189.03	927.53
94.00	751.30	751.30	1071.48
95.00	2070.99	2070.99	970.99
96.00	1696.71	1991.49	1091.49
97.00	318.08	718.08	884.15
98.00	509.14	909.14	938.48
99.00	763.84	527.51	1263.84
100.00	873.39	873.39	712.74

2   Cost

Readings for Chapter 2

Daganzo and Newell (1990) describe a model for handling operations, and examine trade-offs among handling transportation and inventory costs. Section 2.4 covers much of the same material. Blumenfeld, Hall, and Jordan (1985), and Horowitz and Daganzo (1986) examine minimum cost shipping strategies, with random demand and travel times, when a fast and expensive transportation mode can be used to forestall shortages. Part of Section 2.5 is devoted to this subject.

2.1   Initial Remarks

This chapter describes how to account for the various costs arising from a logistics operation; it also introduces related terminology and notation. Although this will be done in the context of a single origin producing identical items[1] for a single consuming destination, the formulas and concepts extend to the more general scenarios examined in the latter chapters of this monograph. Any modifications are described in these chapters. This section presents a framework for the classification of logistics cost; specific cost types will be analyzed in the following sections.

In tracing the path of an item from production to consumption, we see that it must be:

(i)   carried (handled) from the production area to a storage area,
(ii)   held in this area with other items, where they wait for a transportation vehicle,
(iii)   loaded into a transportation vehicle,
(iv)   transported to the destination, and
(v)   unloaded, handled, and held for consumption at the destination.

---

[1] In this monograph we will often call the indivisible units that move over a logistics system, e.g., persons, letters, parcels, etc., "items." When the logistics system handles an infinitely divisible commodity, such as fluids and grain, the term "item" may also be used; in that context it will denote a fixed, and usually small, quantity of the commodity.

These operations incur costs related to *motion* (i.e., overcoming distance) and cost related to *"holding"* (i.e., overcoming time).

Motion costs are classified as either *handling* costs or *transportation* costs. They are very similar; the main difference being the distances transported and the size of the batches moved together. Handling costs include *packaging* (in step *(i)* above); transportation costs include *loading*. Of course, loading is also a handling activity; and if a clear distinction is desired, one could define as a *handling* cost the portion of loading costs that arise *outside* the *transportation vehicle*, and as a transportation cost, the portion that arises inside the vehicle. It is not really crucial that the cost of the specific action be allocated to a "correct" category. What is important is that in the final analysis all costs are included and none are double counted.

Holding costs include *"rent"* costs and *"waiting"* costs. This is not a generally accepted terminology, but it is useful for our purposes. As the name implies, rent costs include the *rent* for the space, *machinery* needed to store the items in place, plus any *maintenance* costs (such as security, utilities, etc.) directly related to the provision of storage space. Waiting costs are meant to capture the cost of delay to the items, including: the opportunity cost of the capital tied up in storage, any value lost while waiting, etc. For a given set of fixed facilities (machinery and space), thus, the rent costs remain fixed, but the waiting costs depend on how the items are processed; i.e., the rent – unlike the total waiting cost per unit time – does not depend on the amount stored. We will examine these four cost categories one by one, and see how they can be quantified. Our goal is to identify which parameters influence the various costs, and the mathematical form of the relationships.

In analyzing these relationships, it is also important to choose how to present them. For example, one could measure transportation cost as: cost per item transported, cost per year, cost per trip, etc. But not all of these representations are valid for analysis. The cost per item can be converted to cost per year if we multiply it by the number of items produced in a year. The cost per item can be converted to cost per trip if we multiply it by the number of items in the transportation vehicle. Two representations are equivalent if the *conversion factor* is *a constant* that does not depend on the decision variables. For example, if we seek the optimal vehicle dispatching frequency that will maximize the yearly profit for a given production level, the desired solution can be found by minimizing the total cost per year – when price and production levels are constant, yearly profit can be related to yearly costs by a known non-increasing function. The same solution could also be obtained by minimizing the average cost per item because the conversion factor, items produced per year, is a constant. The cost per trip, however, would lead to an erroneous solution. In the remainder of this monograph we will assume that the yearly demand for items does not depend on the decision variables, and, therefore, it will be possible to express cost either as a total per unit time or a prorated average per item.

In our discussion we will usually include all the costs incurred by the items from origin to destination regardless of who pays them (the shipper, the carrier, or somebody else). If ownership of the item changes at some point during transportation (e.g., on arrival at the destination), waiting costs at the origin will be paid by the producer, and inventory costs at the destination by the consumer. While one may feel that costs borne by any entity other than our "client" (i.e., the organization whose operation we are trying to optimize) should be ignored, this is shortsighted. Such an optimization would tend to transfer the burden of the operation to entities other than our client (since their costs are not being considered); and as a result, they may be less willing to participate in the operation. If, for example, a producer ships infrequently (which minimizes its own transportation costs) and, as a result, causes large inventories at the destination, the consumer will be less willing to pay the price – and may expect a discount. Such a discount would obviously have to be included in the optimization of the shipping frequency, but it is difficult to quantify. Our expressions automatically include the quantity that the discount would represent – the increased cost to the consumer. Of course, if this is not desired, appropriate terms can be deleted from the expressions; the techniques remain the same.

Let us now turn our attention to the various cost components. Section 2.2 discusses holding costs, Sec. 2.3 transportation costs and Sec. 2.4 handling costs. Section 2.5 explains how uncertainty and random phenomena influence cost accounting.

2.2 Holding Costs

A sufficiently detailed quantitative description of holding costs can be given in the context of a simple scenario with one origin and one destination. Consider the situation depicted in Fig. 2.1, where items are produced and demanded at a constant rate, D'. The four curves of the figure represent the cumulative number of items to have been: (i) produced, (ii) shipped, (iii) received at the destination, and (iv) consumed. We assume that the ordinates of the curves at time zero (when observation began) have been chosen so as to ensure that the vertical separation between any two curves at that time equals the number of items initially observed between the corresponding stations.

Rarely used in the inventory and queueing literature, cumulative count curves such as those depicted in Fig. 2.1 are particularly useful to trace items through consecutive stages. In our case, they conveniently describe in *one* picture how the number of items in *various*, logistic states (waiting for transportation, being transported, and waiting for consumption) change with time. Notice that the number of items waiting for transportation at *any given time* is the vertical separation between curves (i) and (ii) at the corresponding point on the time axis, the number being transported is the vertical separation between curves (ii) and (iii), and the number waiting for consumption is the vertical separation between curves (iii) and (iv).

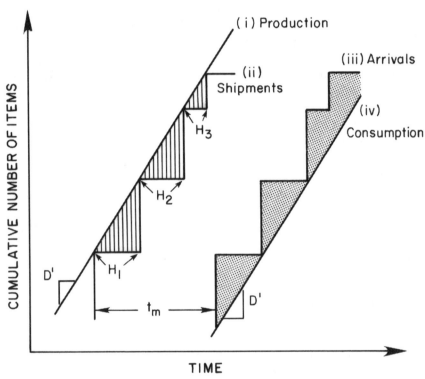

Fig. 2.1   Cumulative item counts at different stages in the logistics operation

Chapter 1 in Newell (1982) shows in detail how various other measures of performance can also be gleaned from these graphs. Of special interest here are horizontal separations between the curves and the intervening areas. When items pass through the system in a "first-in-first-out" order, then the nth item to be counted at each observation station (i, ii, iii, or iv) is the same item; as a result, the horizontal separation between any two curves at ordinate "n" represents the amount of time spent by that item between the corresponding stations. In the figure, thus, t_m represents the transportation time. It should be intuitive that areas between curves represent total amount of wait (in "item-hours") regardless of the order in which items are processed. Thus, the shaded area in the figure represents the number of "item-hours" spent at the origin, and the dotted area represents the number at the destination. It follows that the average horizontal separation between two curves, measured between two points where the curves touch, represents the average time that a typical item spends between the operations represented by the curves. (The average horizontal separation between the curves can be expressed as the ratio of the area, i.e., the total wait, and the vertical separation between the two points, i.e., the number of items processed. Such a ratio is, by definition, the average wait per item.)

In our example, the (constant) separation between the production and consumption curves represents the average "waiting" that an item has to do between production and consumption. This is equal to t_m plus the *maximum* interval (or headway) between successive dispatches, $H_1 = \max\{H_i\}$ (see figure):

$$\overline{wait} = H_1 + t_m. \qquad (2.1a)$$

The room needed for storage at any given location should be proportional to the maximum number of items present at the location. This is represented in Fig. 2.1 by the maximum vertical separations between curves. Because the figure assumes that each vehicle carries all the items that have been produced, the storage area required at the origin is proportional to the maximum headway (otherwise the maximum inventory accumulation would be larger); i.e.:

$$maximum\ accumulation = D'H_1 \qquad (2.1b)$$

The maximum accumulation at the destination is the same as it is at the origin (the reader can verify this from the geometry of the figure, remembering that $H_1 = \max\{H_i\}$).

The expressions for average wait and maximum accumulation can be translated into costs *per item* or *per unit time* using cost conversion factors.

2.2.1 Rent Cost

This is the cost of the space and facilities needed to hold the maximum accumulation; for properly designed systems it should be proportional to the maximum accumulation. The proportionality factor will depend on the size of the items, their storage requirements, and the prevailing rents for space. If the facilities are owned (and not leased), then the purchase cost should increase roughly linearly with size. Thus, one can compute an equivalent rent (based on the amortized investment cost over the life of the facilities) which should still be roughly proportional to the maximum accumulation.

Let c_r be the proportionality constant (in $ per item-year); then

$$rent\ cost/year = c_r\ (maxiumum\ accumulation) \tag{2.2a}$$

and if the demand is constant, Eq. (2.1b) allows us to write:

$$rent\ cost/item = c_r\ (max\ accumulation)/D'=c_rH_1 \tag{2.2b}$$

Note that the rent cost per item is independent of flow (the production and consumption rate D') and proportional to the maximum time between dispatches.

2.2.2 Waiting Cost

Waiting cost, also called inventory cost, is the cost associated with delay to the items. As is commonly done in the inventory literature, it will be captured by the product of the total wait done by all items and a constant, c_i , representing the penalty paid for holding one item for one time unit (usually a year). Thus,

$$waiting\ cost/year = c_i\left(total\ wait\ per\ year\right)$$

and

$$waiting\ cost/item = c_i\left(average\ wait/item\right).$$

Because the above expressions implicitly value all the item-hours equally, caution must be exercised when the penalty depends on: (i) the time of day, week, or year when the wait occurs, and (ii) how long a specific item has already waited. For the example in Fig. 2.1, the waiting cost is:

$$waiting\ cost/year = c_i\left[D'(H_i + t_m)\right]$$
$$= (c_iD'H_1) + (c_iD't_m)$$
(2.3a)

$$waiting\ cost/item\quad = c_i(H_1 + t_m)$$
$$= (c_iH_1) + (c_it_m)$$
(2.3b)

The left side of Equation (2.3a) assumes that the time unit is one year. The term in brackets represents the average accumulation of inventory in the system (the vertical separation between the production and consumption curves of Fig. 2.1). As we shall see, it is usually convenient to group the terms associated with H_1 in Eqs. (2.2b) and (2.3b), by defining a stationary holding cost per item-day $c_h = c_r + c_i$.

For problems in which the inventory at the destination can be ignored (e.g. for the transportation of people in many cases) the average wait added to t_m should be computed for the shaded area in Fig. 2.1. The result, a value somewhere in between $\frac{1}{2}\,c_i\bar{H}$ and $\frac{1}{2}c_iH_1$, is no longer a function of H_1 alone.

If we were shipping people, c_i would represent the "value of time". When shipping freight, this constant would include the opportunity cost of the capital tied up in holding an item for one time unit. (If π denotes the "value" of an item, and i an agreed upon discount rate, then the opportunity cost is πi). For perishable items, and items exposed to loss and damage, c_i should also include any value losses arising from time spent in the system. The constant, c_i is hard to determine precisely. We don't know people's value of time accurately and, as is well known in economics, it is hard to pinpoint "i". Furthermore, in most cases even the value of the items themselves is hard to measure.

Suppose that an item costs π_0 dollars to produce but it is sold for π_1 dollars ($\pi_1 \gg \pi_0$). Which of these two values should be used for inventory calculations? The answer depends on market conditions. If the demand is fixed, a reduction in inventory allows the production to be slowed (temporarily only) until the new lower inventory levels are reached (see Fig. 2.2). If the wait is reduced by Δ units, the production of $D'\Delta$ items can be avoided. The resulting one-time savings can be amortized over the life of the operation to yield a cost savings per unit time which is proportional to $D'\Delta\pi_0$. This is the same as saying that c_i is proportional to π_0 .

If on the other hand the market could absorb everything that is produced, one could then sell the extra $D'\Delta$ items in inventory while keeping the production rate constant and the amortized extra revenue per unit time would be proportional to $D'\Delta\pi_1$. This means that c_i would be proportional to π_1.

In practice, one often finds that even π_0 and π_1 are not known; this often happens when the items are components consumed within the firm as part of a multi-plant production process. Accounting systems are typically rigged to track the overall costs of production according to broad categories (e.g., labor, depreciation, etc.) but the costs are not prorated to the different components that are produced.

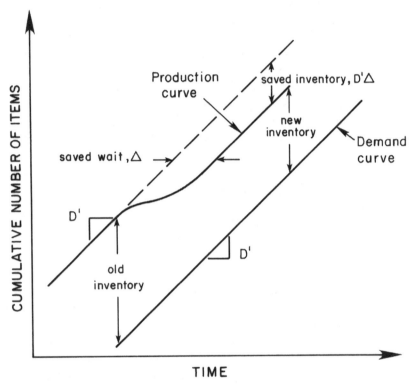

Fig. 2.2   Inventory effect of a temporary reduction in the production rate

In other cases the product can be acquired at different prices from different producers, so π_0 is not fixed. Then, the relevant price used for decision-making is not necessarily the average. For example, if a producer can secure limited supplies of items both for a low price ($\pi_0 = \pi$) and unlimited supply at a higher price ($\pi_0 = \pi' > \pi$), it will try to meet as much of its demand with the cheap items. If the demand rate comfortably exceeds the capacity of the cheap supplier, further increases in the rate would be satisfied at cost π'.Thus, this high value and not an average would be the relevant cost for an analysis of possible market expansions. Clearly, careful consideration is often necessary in determining something as seemingly basic as the "cost of goods sold".

Finally, the value of c_i that one would use in expressions such as Eqs. (2.3) should also reflect any indirect costs of delay to other aspects of the overall operation such as the effect of inventories on quality. These effects may be hard to quantify, but must be considered. Conventional wisdom indicates that large inventories lower quality because their existence reduces the incentive to eliminate defects *at the origin* – after all, items found to be defective can be replaced from the existing stock. Without this incentive, the quality of all the items (even those that are not defective) may suffer.

The value of c_i can change by many orders of magnitude, depending on what is being transported. For people c_i should be on the order of $10 per hour so that a bus load of 30 people would cost between 10^2 and 10^3 dollars per hour. A truck carrying 20,000 lbs of goods costing on the order of $1 per pound (which would be typical of groceries, machinery, etc.) would contain cargo valued at $20,000. Amortized at 10 percent for a (2,000 hour) year, the cargo costs on the order of 10^0 per hour. Cheaper and lighter cargoes can result in even lower costs. These "back-of-the-envelope" calculations illustrate that while it may be difficult to define c_i very precisely in any specific application, it should be possible to estimate its order of magnitude. Fortunately, rough estimates often are all that is needed. As we shall see, the structure of a logistic system depends on the order of magnitude of c_i, but it is not very sensitive to small changes in c_i.

Before turning our attention to motion costs, let us introduce some terminology to identify the two terms, $(c_i H_1)$ and $(c_i t_m)$, of Eq. (2.3b). The first term, which depends on the maximum dispatching headway and arises when the items are stationary, will be called the "stationary inventory cost." The other component $(c_i t_m)$, which arises while the items are moving and is independent of the dispatching headways, will be termed "pipeline inventory cost."

The following two sections discuss motion costs. Transportation costs are addressed first.

2.3   Transportation Costs

We continue with the one-origin/one-destination situation that was depicted in Fig. 2.1. If one uses a public carrier to transport the items from the origin to the destination, the total cost per year will be the sum of the costs of each individual shipment. Published rates increase roughly linearly with shipment size. (The rates increase in steps, but the overall slope is approximately constant for wide ranges of shipment sizes.) The mathematical relationship is:

$$shipment\ cost \approx c_f + c_v v. \tag{2.4a}$$

where v is the shipment size, c_f is a fixed cost per shipment that should include things such as driver wages, and c_v is the rate at which the variable cost per shipment increases size, e.g. due to increased fuel consumption. The cost for shipping a sequence $\{v_i\}$ of n shipments (i = 1, ... ,n) totaling V items ($V = \Sigma_i\ v_i$) is thus:

$$cost\ for\ n\ shipments = \sum_{i=1}^{n} c_f + c_v v_i = c_f n + c_v V. \tag{2.4b}$$

The total cost only depends on the number of shipments, *regardless of what they contain and when they happen*, and the total number of items shipped. The cost per item, thus, decreases with the average shipment size, $\overline{v}$

$$transportation\ cost/item = c_f \left(\frac{n}{V} \right) + c_v = c_f \left(\frac{1}{\overline{v}} \right) + c_v. \tag{2.5a}$$

These economies of scale arise because all the items in a shipment share the fixed cost, c_f.

For our simple problem with one origin and one destination, the only decision variable appearing in Eq. (2.5a) is n (or $\overline{v}$); thus, the variable cost should not influence shipping decisions. We will not eliminate it from our expression, though, because c_v is not a constant for more complicated problems. (As we shall see, c_v depends on distance; and for problems with many origins and destinations, the distance traveled is not fixed.)

2.3.1   Relationship to Headways

Like inventory and holding costs, the cost of transportation depends on the dispatching headways. The relationship is:

$$Transportation\ cost/item = c_f / \left(D'\overline{H} \right) + c_v \tag{2.5b}$$

because $V = \Sigma_i\ D'H_i = D'\overline{H}\ n$; i.e., $\overline{v} = D'\overline{H}$. The transportation cost decreases with the *average* headway, unlike holding costs which increased with the *maximum* headway. Notice as well that for a given number of shipments, and thus a given average headway, the transportation cost is independent of the specific headways. Hence, *shipments should be spread as regularly as practicable* to reduce the maximum headway and the associated holding cost. If headways can be maintained constant, $H_i = H$, then both holding and transportation costs are functions of H.

2.3.2 Relationship to Distance

As an aside that will become important for multiple origin and/or multiple destination logistic problems, we examine the relationship between transportation cost and distance.

Rate books reveal that c_f and c_v depend mainly on distance; the precise location or origins and destinations also influences these costs but to a lesser extent. The relationships are well approximated by linearly increasing functions of distance, d:

$$c_f = c_s + c_d d \ \ and \ \ c_v = c'_s + c'_d d.$$

The interpretation of these four new constants appearing in the right side of these expressions is easier when the above expressions are substituted for c_f and c_v in Eq. (2.4b). The cost for n shipments totaling V items, when the origin and destination are d distance units apart, can be broken up in four terms as follows:

$$\begin{pmatrix} cost \ for \ n \\ shipments \end{pmatrix} \approx c_s n + c_d nd + c'_s V + c'_d Vd. \tag{2.5c}$$

The first constant, c_s , is the cost attributable to each trip, regardless of distance and shipment composition; it includes the cost of stopping the vehicle and having it sit idle while it is being loaded and unloaded. Think of it as the fixed cost of *stopping* "c_s", independent of what is being loaded and unloaded. The second constant, c_d , is the cost attributable to each incremental vehicle-mile. It is the vehicle cost (including the driver) for each mile traveled regardless of the vehicle's contents; i.e., the cost of *distance*, "c_d."

The third constant, c'_s , represents the added cost of carrying an extra item. It represents a penalty for delaying the vehicle while loading and unloading the item, as well as the cost of handling the item within the vehicle. (Handling costs outside the vehicle will be considered in Section 2.4.).

The fourth constant is the cost attributable to each incremental item-mile. It can be viewed as the marginal wear and tear and operating cost *per mile* for each extra item carried. This constant, and the fourth term as a whole, should be small compared with the second term (since the cost of a vehicle-mile is relatively independent of a vehicle's contents); it will normally be ignored.

If, instead of a single destination, the vehicle carried the items picked up at the origin to several destinations, making in the process n_s delivery stops, Eq. (2.5c) would likely have to be modified slightly. Logically, rates must reflect the additional delay-cost for the extra stops. However, because not much else changes (the vehicle travels the same distance and carries the same number of items), one would expect only the first term of Eq. (2.5c) to change. Although not verified experimentally, it seems reasonable to expect it will increase proportionately to the number of stops ($1 + n_s$). Accordingly, if we redefine c_s to be the fixed cost *per* stop, then the cost of making n shipments is

$$\begin{pmatrix} cost\ for\ n \\ shipments \end{pmatrix} \approx c_s(1+n_s)n + c_d\,n\,d + c'_s\,V, \qquad (2.5d)$$

where the fourth term of Eq. (2.5c) has been neglected.

Whether Eq. (2.5d) matches actual rates when $n_s > 1$ is an open question. Multiple stops, however, are normally made as part of exclusive service agreements between shippers and carriers, which should reflect the carrier's actual operating costs; in that case, Eq. (2.5d) seems justified. That carrier cost (or the shipper cost if it uses its own vehicle fleet) is well approximated by Eq. (2.5d) should be intuitive. Drivers' wages should be proportional to the total vehicle-time for all the trips. Because vehicle depreciation cost (overhead) is proportional to fleet size, i.e., the number of vehicle-years needed per year if the demand for vehicles is not seasonal, overhead can be prorated to the tasks of a year on a total vehicle-time basis. Thus, the sum of overhead and driver wages is proportional to the total vehicle-time for the n shipments. Other vehicle operating costs should be proportional to the total number of *moving* vehicle-hours. Because both the total time and the time in motion are linear functions of the vehicle-miles traveled nd , the number of stops n($1 + n_s$), and the total amount of freight hauled V , the total cost should be roughly linear in these variables; i.e., Eq. (2.5d) is a good approximation for the carrier cost.

On dividing Eq. (2.5d) by V, the average cost per item is obtained:

$$cost/item \approx c_s \frac{1+n_s}{\overline{v}} + c_d \frac{d}{\overline{v}} + c'_s.$$

As a function of the average headway, the costs per item and per unit time are:

$$cost/item \approx c_s \left(\frac{1+n_s}{\overline{D'H}} \right) + c_d \left(\frac{d}{\overline{D'H}} \right) + c'_s. \qquad (2.5e)$$

$$cost/time \approx c_s \left(\frac{1+n_s}{\overline{H}} \right) + c_d \left(\frac{d}{\overline{H}} \right) + c'_s D'. \qquad (2.5f)$$

Although Eqs. (2.5e and 2.5f) do not show a dependence on the individual headways, we should recognize that irregular schedules may require slightly larger cost coefficients if the shipper exclusively uses its own private fleet.

This happens because the fleet size needed is dictated by the operation of the system during time periods with the largest numbers of dispatches, with the result that fleet size costs are more closely related to the minimum headway than to the average. An extensive discussion of this issue for a problem with variable demand can be found in Hurdle (1973a) and (1973b); see also Du (1993). Fleet size considerations, thus, provide a *second* incentive to keep transportation schedules as regular as possible.

Finally, note that the in-vehicle time of a typical item, t_m, is also a linear function of distance, d, and number of stops, n_s. This observation will become important later when vehicle routing is a decision variable.

2.3.3 Relationship to Size; Capacity Restrictions

Let us now return to the single origin and single destination situation of Fig. 2.1. So far, we have ignored the possibility of sending *very* large shipments; shipments that would not fit in the largest vehicles on the road. If one were to plot the cost per shipment versus shipment size for a range extending beyond this maximum, v_{max}, for a firm that owns its own vehicles, one would likely find a graph as the one shown in Fig. 2.3. Whenever the shipment size reaches and exceeds a multiple of v_{max} a new vehicle needs to be dispatched with a resulting jump in cost. The steps of Fig. 2.3 should be rather flat (with $c_v v_{max} \ll c_f$) since the cost of operating a vehicle is rather insensitive to what it contains. Whether or not it is exactly as shown in Fig. 2.3, the transportation cost per shipment function, $f_t(v)$, should be "subadditive;" i.e., it must satisfy: $f_t(x_1 + x_2) \le f_t(x_1) + f_t(x_2)$ for any x_1, $x_2 \ge 0$. This property is to be expected because one should not be able to reduce the cost of a shipment by shipping it in parts (see Problem 2.3).

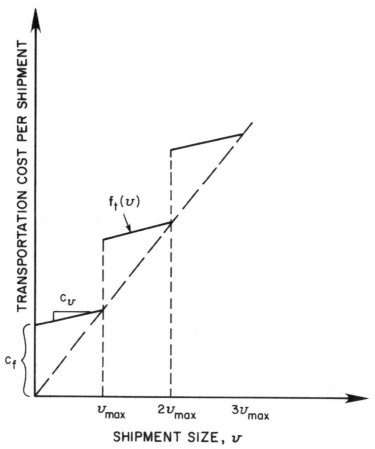

Fig. 2.3   Relationship between transportation cost per shipment and shipment size

For most problems, though, one only needs to consider the linear part of f_t between 0 and v_{max}, as shipments larger than v_{max} are not economical. This can be easily seen if handling costs can be ignored (e.g., if the handling cost per item is a constant, independent of shipment size) by examining the sum of the average *holding* and *motion* costs per item as a function of shipment size. Figure 2.4 plots the average transportation cost per item as would be obtained from Fig. 2.3. The figure also plots *the negative* of the holding costs as a function of shipment size. (We are assuming here that headways are regular, $\bar{H} = H_1 = H$; and we are using Eqs. (2.2b) and (2.3b) with $H_1 = H = v/D'$. Recall that c_h is the stationary holding cost per item-day, $c_h = c_i + c_r$).

The optimal shipment size is the value of v for which the vertical separation between the two curves of Fig. 2.4 is minimum. Clearly, the point can be identified by sliding the "waiting" curve upwards until it first touches the transportation curve. This can only happen either at point P of the figure (where $v = v_{max}$), or else at a point $v < v_{max}$, if the line is sufficiently steep. For most problems, thus, one can ignore the behavior of the transportation curve for $v > v_{max}$, if one remembers to abide by the constraint: $v \leq v_{max}$.

Analytically, the optimal shipment size of Fig. 2.4 is the solution of the following problem:

$$(EOQ): \min \left\{ Av + \frac{B}{v} \right\} \ s.t.: \ v \leq v_{max} \ .$$

where

$$A = c_h / D', and \ B = c_f \ .$$

This is the well known "lot size" or "economic order quantity (EOQ)" model of the inventory control literature (Welch, 1956; Arrow et al., 1958) whose roots can be traced to the pioneering work of F.W. Harris in the early part of this century (Harris, 1913a and 1913b). Erlenkotter (1990) describes these works in a historical context.

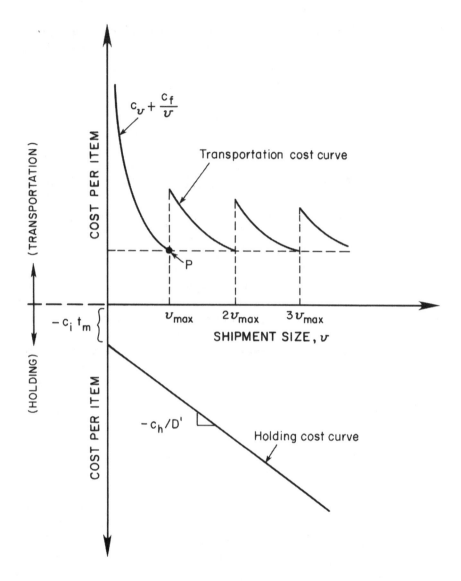

Fig. 2.4 Transportation and holding cost (per item) as functions of shipment size

2.3.4   Relationship to Size: Multiple Transportation Modes

We have already seen that shipment cost increases approximately linearly with size (Eq. 2.4), and that this is likely to be true for fairly broad ranges of shipment sizes. This qualification was made because if shipment size varies by a large amount, it may be cost effective to change transportation modes.

While some shipping modes, such as mail, exhibit a low fixed cost per shipment and a high cost per item, others may be the opposite. Fig. 2.5 shows three such curves. Note that the best mode depends on the shipment size; as it grows, one tends to favor the modes with lower variable cost and higher fixed cost. (In comparing modes, the vehicle cost should include the fixed pipeline inventory cost per item, $c_i t_m$; faster modes may be preferred for valuable items.)

Fig. 2.5 displays the transportation cost that results if one ships everything by the cheapest mode – the lower envelope of the three cost curves. If, as shown, cost increases at a decreasing rate for each mode, then the lower envelope also increases at a decreasing rate. Like the cost curves for the specific modes, the shipment cost by the best mode is then increasing and concave, and therefore subadditive; this shows that cost cannot be reduced further by breaking the shipment into parts. The lower envelope is optimal.

If the individual modal component curves are merely subadditive, e.g., they exhibit jumps as in Fig. 2.3, then the lower envelope is not necessarily optimal, or subadditive. In this case, costs can sometimes be reduced by breaking a shipment into parts and sending it by different modes. For example, if the cost parameters of two modes with $v_{max} = 1$ were ($c_f = 1$, $c_v = 0$) and ($c_f = 0$, $c_v = 1.5$), then the single-mode shipment cost for $v = 1.1$ would be either 2 or 1.65; i.e., 1.65 by the best mode. But this is not optimal. The optimum is achieved by sending a one-unit shipment with the first mode (cost = 1) and the remainder with the second mode (cost = 0.165). If shipments can be allocated to the modes in an optimal way and the modal cost curves are subadditive and increasing, then the overall cost curve can itself be shown to be subadditive and increasing (see problem 2.4).

If the shipper operates its own vehicle fleet, the curves of Fig. 2.5 could represent different vehicle types, and the figure would then indicate the most economical vehicle type for the particular shipment size. Because such a choice is not as flexible as a choice of public carriers (i.e. modes), shippers do not change the vehicle fleet often. When the choice of vehicle type is not an issue, then the appropriate (linear) component curve should be used to evaluate transportation cost.

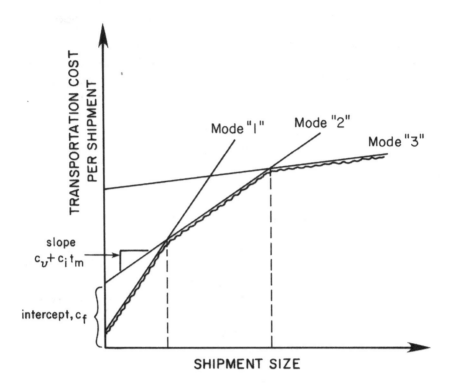

Fig. 2.5   Relationship between shipment cost and size for various transportation modes

2.4 Handling Costs

Handling costs include loading individual items onto a "container", moving the container to the transportation vehicle threshold, and reversing these operations at the destination. The container can be a *box* or a *pallet,* or if the items are large enough, nothing at all. We examine here the cost of handling a shipment of size, v .

If the items are handled individually, the handling cost per shipment should be proportional to v , so that

$$handling\ cost \approx c'_s\ v.$$

If the items are small, it is not economical to move them individually; instead they can be moved on "handling vehicles" such as pallets. Clearly, the handling cost should have a similar form as the transportation function, since items are being transported within a compound. If the batch is smaller than one pallet the cost of handling it should therefore be:

$$handling\ cost/batch \approx c'_f + c'_v\ v. \tag{2.6}$$

The constant c'_f represents the (fixed) cost of moving the pallet regardless of what it contains, including the forklift driver's wages, plus the forklift's depreciation and operating cost. The constant c'_v captures the cost, accounting for both labor and capital, of loading one item on the pallet. If v is larger than the maximum number of items that fit on a pallet, v'_{max}, then the handling cost function per shipment, $f_h(v)$ will still be a scaled down version of the transportation function, as in Fig. 2.6.

At the destination, the handling cost function will be analogous, possibly with different c_f' and c_v' but the same v'_{max}. As a result, the combined handling cost for the shipment at both ends of the trip should still have the form of Fig. 2.6, and should obey Eq. (2.6) if $v \leq v'_{max}$.[2]

One could compare the cost of moving items individually and moving them in pallets. But if more than one item fits on a pallet, it will usually be cheaper to move them in pallets.

---

[2] Although we have used the words pallet and forklift repeatedly, we stress here that Eq. (2.6) also applies to other container-filling methodologies that do not use forklifts; e.g., to the "bucket-brigade" method of order-picking using passive conveyors described in Bartholdi and Eisenstein (1996). In these cases, one just needs to make sure that the constants c'_f and c'_v are representative of the actual operation.

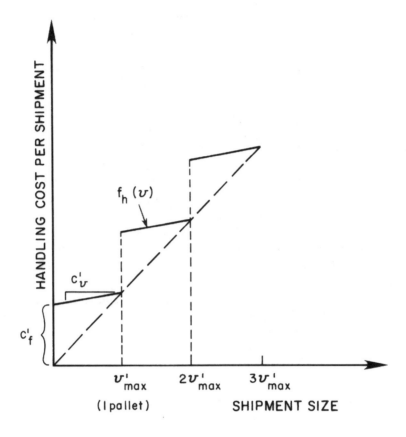

Fig. 2.6   Handling cost per shipment as a function of shipment size

2.4.1 Motion cost

Figure 2.7 depicts the sum of transportation plus handling costs for v'_{max} $\ll v_{max}$. The function, $f_m = f_t + f_h$, is still subadditive and increasing. (See problem 2.4.)

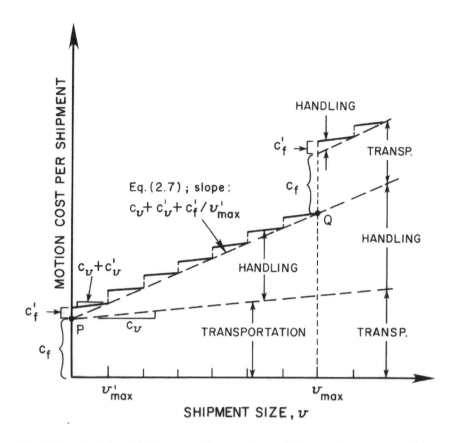

Fig. 2.7   Relationship between shipment size and the combined cost per shipment of transportation and handling

Note that to within an error of c'_f , the motion cost per shipment, $f_m(v)$,
can be approximated by line $\overline{PQ}$ of the figure, which is a lower bound:

$$f_m(v) \approx c_f + \left(c_v + c'_v + \frac{c'_f}{v'_{max}} \right) v . \qquad (2.7)$$

This indicates that handling costs can be subsumed in the transportation
cost function, Eq. (2.4a), with a suitable definition for the fixed and variable cost: [3]

---

[3] If $v < v'_{max}$ it is better to use: $c''_f = c_f + c'_f$ and $c''_v = c_v + c'_v$

$$c''_f = c_f \quad and \quad c''_v = c_v + c'_v + \frac{c'_f}{v'_{max}}.$$

The expression for the variable motion cost per item, c''_v , is intuitive; in addition to the variable transportation and handling costs per item, c_v and c'_v , it includes each item's prorated share of the fixed cost per pallet, c'_f / v'_{max} .

2.4.2   The Lot Size Trade-Off with Handling Costs

If we prorate the cost of a shipment to the items that it contains, we can construct a figure, analogous to Fig. 2.4, which can be used to determine the optimal size of the shipments. Figure 2.8 is not extended beyond v_{max} , since larger shipments continue to be undesirable. Note from the figure that if the waiting cost curve is pushed upwards, the first point of contact is either $v < v'_{max}$ (if the waiting cost curve is very steep), or else it is likely to be an integer multiple of v'_{max} . (This is not always the case, but very little is lost by assuming that it is – Daganzo and Newell, 1987). Because the lower bound from Eq. 2.7 is exact when v is an integer multiple of v'_{max} , one could use it instead of the exact (scalloped) curve while restricting v to be a multiple of v'_{max} . Except for the variable cost coefficient, c''_v , this equation matches Eq. (2.4a) , and we saw already that variable costs do not influence the optimal shipment size. Thus, if shipment size is restricted to be an integer multiple of v'_{max} , *the optimal shipment size is independent of handling costs.*

We now examine the consequences of relaxing this restriction. If the optimal shipment size, v^* , is greater than one pallet, we see from Fig. 2.8 that allowing v to differ from a multiple of a pallet cannot improve things appreciably. In the most favorable case the cost savings can be shown to be about one tenth of c_f/v'_{max} , with much smaller savings in other cases; see problem 2.5. Thus, even without the restriction, one can safely ignore handling costs in determining shipment size.

If, on the other hand, v^* is smaller than one pallet, then handling costs should be considered; there may be a significant difference between $f_m(v)$ and its lower bound (see Fig. 2.8 and the previous footnote). If $c'_f \gg c_f$, then the optimal shipment size may be noticeably larger than if handling costs had been ignored.

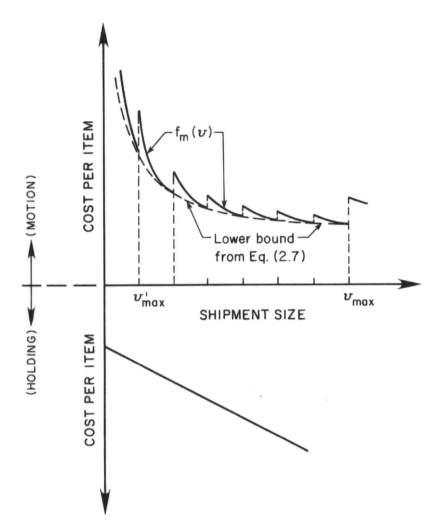

Fig. 2.8   Motion and holding costs (per item) as functions of shipment size

In summary, the following simple recipe can be used: If economic shipment sizes are likely to be larger than a pallet, ignore handling costs in the decision; but if shipment sizes are smaller than a pallet, then include the fixed cost of handling a pallet as part of the fixed cost per shipment and select the shipment size which is the minimum of problem "EOQ" with $A = c_h/D'$ and $B = c''_f$.

More complicated motion curves would arise if items had to be put into boxes, which could be put onto pallets, which would then travel on trucks. Because the relationship of boxes to pallets is analogous to the relationship between pallets and trucks, the additional handling step would be reflected by a second set of scallops on Fig. 2.8. The selection of an optimal shipment size would be affected by this second set of scales in a similar way: *the cost of moving and filling boxes can be ignored if the optimal shipment size is larger than a box*; otherwise, the fixed cost per shipment, c''_f, should include the fixed cost of moving one box including opening and closing it, but not the cost of filling it.

With a properly defined c''_f, the optimal shipment size should still follow from the solution of the "EOQ" problem:

$$(EOQ) \min \left\{ Av + \frac{B}{v} + C \right\} ; v \le v_{max} , \qquad (2.8a)$$

where:

$$A = c_h/D' , \ B = c''_f , and \ C = c_i \, t_m + c''_v . \qquad (2.8b)$$

Note that c''_v should include any handling costs (per item) not included in c''_f, and that the minimum of Eq. (2.8a) is unaffected by c''_v since C is an additive constant in Eq. (2.8a). Also remember that if the minimum of (2.8a) is greater than one box (or pallet) the shipment size should then be rounded to the nearest box (or pallet); the cost, however, remains close to the minimum of (2.8a) without rounding.

2.5   Stochastic Effects

We have assumed in our discussion of cost that the transportation travel time and the production and consumption rates are constant. These assumptions can be violated in two ways. The production and demand rates (and the travel time perhaps as well) may vary over time in a predictable manner, and also unpredictably. Predictable variations such as seasonal trends and day of the week effects will be examined in Chapter 3; optimal decisions can be found because costs can be predicted.

Unpredictable variations are another matter and are examined here; they require additional inventories, and may also increase transportation cost. Continuing with the single origin and single destination model, for the rest of this section we assume that production is driven by consumption. That is, the destination requests deliveries so that its inventory level can sustain at all times the demand that is anticipated. With inherently unpredictable demand and travel times, however, it is no longer possible to time the shipments so they arrive just as the stock at the destination is running out, as with the first two shipments of Figure 2.1. Stochastic variations are the subject of much attention in the inventory control literature, where the objective is to determine optional levels of "safety stock" and reorder "trigger points" (see Peterson and Silver (1979) or Zipkin (2000) for example).

These stochastic phenomena complicate matters, but in many cases the *added* holding plus motion costs (per item) that arise due to randomness can be shown to be known linear functions of either v or 1/v, and in other cases completely independent of v . This is fortunate because the added costs can then be captured by a deterministic EOQ model, Eq. (2.8a), where some of the constants ("A", "B", and "C") have been increased.

2.5.1 Stochastic Effects Using Public Carriers

Newell, in some unpublished notes, has pointed out that if transportation is reliable enough to ensure that shipments arrive at the destination in the order in which they were requested, then the added cost due to randomness is constant. As a result, the demand and travel time uncertainty should influence neither the frequency of dispatching nor the average lot size.

A common ordering strategy uses a trigger point v_0 as follows: whenever the inventory on hand *plus* the number of items on back order equals v_0 , a shipment of size v is requested.[4] The reorder headways for this strategy vary because the demand varies, but the shipment sizes remain constant.

Let us assume that the demand arrival process can be approximated by a diffusion process with rate D' (items per unit time) and index of dispersion γ (items).[5] The index of dispersion represents the variance to mean ratio of the number of items to have arrived in one time unit. (Note that if items are measured by a physical quantity such as tons, cubic feet, etc., γ shares these units.) A suitable choice of γ approximates most of the processes examined in the inventory literature. Let us also assume that the lead time, T_ℓ, (the time between order placement and receiving) has mean t_ℓ and standard deviation σ_ℓ. (The lead time should be close to the average transportation time, t_m , if the origin can keep up with the requests; but this assumption is not needed here.)

---

[4] These strategies are called "(s,S)" in the inventory literature; see e.g., Peterson and Silver (1979); Zipkin (2000).

[5] For a diffusion process, the number of arrivals in any time interval is a normal random variable, with mean and variance proportional to the duration of the interval, and independent of the arrivals in non-overlapping intervals. Newell (1982) proposes to approximate queuing and inventory phenomena with diffusion processes.

If one desires to avoid stock-outs, the trigger point, v_0, should be large enough to ensure that no stock-out occurs immediately before the arrival of an order. The best way of exercising this policy can be found with the help of the three curves in Fig. 2.9, relating time to the cumulative number of items that have been: (i) ordered, (ii) received, and (iii) consumed at the destination. The dashed lines in the figure represent the portion of the curves that is not yet known at time "NOW". A request for a shipment is depicted immediately after time "NOW" since at that time the sum of the inventory on hand and the back orders is shown to be v_0. Because all the back orders are sure to have arrived before the new order, it is clear from the figure that a stock-out will be averted immediately before the new order arrives if the future consumption until the new order arrives (segment $\overline{PQ}$ in Fig. 2.9) does not exceed the inventory currently on hand plus the back orders, v_0.

This condition can be expressed probabilistically if we recognize that, conditional on the lead time, T_ℓ, $\overline{PQ}$ is normal with mean $D'T_\ell$ and variance $D'\gamma T_\ell$. The unconditional first two moments of $\overline{PQ}$ are thus:

$$E\left(\overline{PQ}\right) = D't_\ell$$
$$\mathrm{var}\left(\overline{PQ}\right) = D'^2\,\sigma_\ell^2 + D'\gamma t_\ell.$$

If the trigger point, v_0, is chosen several standard deviations greater than $D't_\ell$, stock-outs will be rare. The precise value of v_0 is not important for our analysis (it is a function of D', γ, t_ℓ, σ_ℓ, and nothing else); what is important is that, as the figure clearly indicates, the contribution of v toward the maximum and average accumulation is insensitive to v_0. This would, in fact, be the case even if v_0 were chosen in a more involved manner (e.g. recognizing the distribution of T_ℓ). Existing methods for selecting trigger points and shipment sizes (Peterson and Silver, 1977, Zipkin, 2000) exploit this insensitivity.

In order to choose the optimal v , the motion and inventory costs must be balanced, as shown in prior sections. In the long run, the motion costs with and without stochastic effects are the same because the same number of shipments are sent in both cases (D'/v shipments per unit time), but the holding costs are larger with stochastic phenomena. The maximum number of items present at the destination will certainly occur after the arrival of an order. As shown in the figure, for a typical order, this number is:

$$v_0 + v - \overline{PQ},$$

which is largest when $\overline{PQ}$ is as small as possible. The term $(v_0 - \overline{PQ})$ represents the contribution of randomness toward higher inventories; but the term is not dependent on our decision variable, v . Thus, except for an additive constant, the holding costs are as in the deterministic case; the optimal shipment size remains the same.

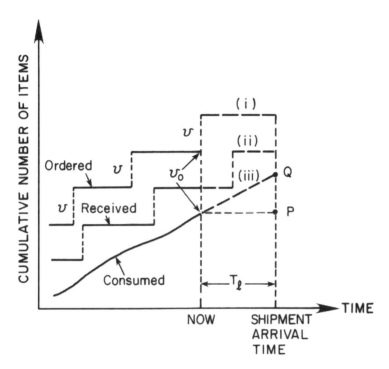

Fig. 2.9   Evolution over time of the cumulative number of items ordered, received and consumed for a simple trigger point strategy

For clarity, the inventory at the origin was ignored in the foregoing discussion. Yet, the irregular way in which orders are placed will undoubtedly raise inventory and production costs at the origin. These effects, however, are shown below to be largely independent of v (their actual magnitude depends on how frequently the production is adjusted) and, thus, should not influence shipping decisions.

If, as is usual, there is an incentive to maintain a steady production rate, then one would set it at a value D'_p , slightly greater than D' to ensure that the overall demand can be met in the long run. Although inventories at the origin would then tend to grow with time, every once in a while (every many reorders, presumably) the production process could be interrupted for a while to allow the demand to catch up with the cumulative number of items produced. The frequency of these stoppages would depend on production and inventory cost considerations.

A simple strategy would stop production whenever the inventory at the origin (after a shipment) reaches a critical value, v_1 , and would resume it (also after a shipment) when the inventory dips below another value, v_2; see Fig. 2.10. The maximum inventory is therefore: $v_1 + v$, and the average inventory: $1/2(v_1 + v_2 + v)$. The cost of production should be a function only of D'_p and the duration of the on and off periods. The on and off periods, however, only depend on v_1 , v_2 and on the statistical properties of the smooth curve tangent to the crests of the orders sent curve (see Fig. 2.10). On a scale large compared with v , this curve shares the statistical properties of the demand curve *which do not depend on v*. Therefore, the optimal production decisions (i.e., the choices of v_1, v_2 , and D'_p) do not depend on v.

As before, the inventory (maximum and average) can be decomposed into a portion that is proportional to v (represented by the shaded area in Fig. 2.10) and independent of the production strategy, and a remaining portion which is influenced by the production scheme and is independent of v . Thus, the extra production and inventory costs arising at both the origin and the destination due to the unpredictability of demand are largely independent of v . They can be ignored when determining the optimal shipment size.

The foregoing discussion is not an exception; stochastic effects can be captured within the scope of a deterministic EOQ model in other situations as well. Problem 2.6 discusses the use of a private vehicle fleet, and the following subsection considers an operation where two different transportation modes are used.

2.5.2 Stochastic Effects Using Two Shipping Modes

It has been assumed so far that stock-outs are avoided by holding inventories large enough to absorb fluctuations in demand and in the transportation lead time. In some instances, if a second, much more expensive, shipping mode is available for expediting shipments, the total costs may be reduced by expediting small shipments at critical times. In these instances the optimal lot size v is also the result of an EOQ trade-off, although the trigger point decision is no longer independent of the shipment size decision. For the following discussion it is assumed that the expedite mode is so fast that its lead time can be ignored.

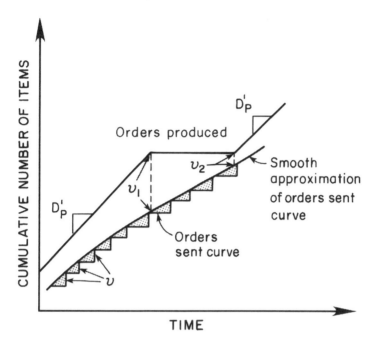

Fig. 2.10   Inventory effect of production and transportation decisions

Most of the time the expedite mode lies in wait, and the system operates as if the primary mode was the only mode (see Fig. 2.9). The trigger point v_0, however, does not have to be chosen as conservatively as before, because when a stock-out is imminent enough items can be sent by the premium mode to avoid it.

The analysis is simple. If, as is commonly the case, the time between reorders is large compared with the primary mode's lead time (i.e., so that when the trigger point, v_0, is reached there aren't any unfilled orders) then the probability that some items have to be expedited in the time between ordering and receiving a lot (of size v) does not depend on v . It is a decreasing function of v_0, approaching zero when $(v_0 - E(\overline{PQ}))^2 \gg$ var($\overline{PQ}$).

The exact form of the expected amount expedited per *regular* shipment will depend on the strategy used for choosing the expedited lot sizes. (Although these could be fixed, if possible they should be chosen just large enough to meet demand until the regular order arrives). In any case, the expected amount expedited per regular shipment will also be a decreasing function of v_0 , $f(v_0)$. Assuming that the cost per item expedited is a constant, c_e , we find that the expected expediting cost per regular shipment is: $c_e f(v_0)$. The moving cost per item is as a result:

$$\binom{moving\ cost}{per\ item} = \frac{c_f + c_v v + c_e f(v_0)}{v + f(v_0)} = \frac{c_f + (c_e - c_v)f(v_0)}{v + f(v_0)} + c_v .$$

The maximum inventory still occurs when $\overline{PQ}$ is as small as possible, and remains: $v + v_0 - \overline{PQ}$; the total cost per item is thus:

$$cost/item = c_v + \frac{c_f + (c_e - c_v)f(v_0)}{v + f(v_0)} + c_h\left(v + v_0 - \overline{PQ}\right)/D' .$$

For a *given* v_0 , if we think of the expected amount shipped by both modes with every regular shipment, $v' = [v + f(v_0)]$ as the "lot size," the equation is still of the EOQ form (2.8a), where the fixed moving cost has been increased to include the expected cost of expediting, $(c_e - c_v)f(v_0)$:

$$cost/item = \left\{ c_v + c_h \left[v_0 - \overline{PQ} - f(v_0) \right]/D' \right\}$$
$$+ \frac{c_f + (c_e - c_v)f(v_0)}{v'} + c_h\ v'/D' .$$

Unlike in the previous case, though, the trigger point v_0 should not be chosen independently of v. If v is large so that shipments are infrequent, expediting a significant amount of freight with the average shipment only increases the moving costs marginally. But if v is small, the penalty for expediting is paid more often; it may be more efficient to increase v_0.

Suggested Exercises

2.1 Prove that if, as depicted in Fig. 2.1, the production and consumption rates are constant and shipments always carry all the production that has accumulated prior to their departure, then the maximum accumulation at the destination equals the maximum accumulation at the origin.

2.2 Use multiple regression analysis to validate Eq. (2.5c) using a recent book of rate tables.

2.3 Prove that the graph of Fig. 2.3 depicts a subadditive function.

2.4 Prove that the following functions are subadditive: (i) any positive, increasing and concave function defined for $x \geq 0$, (ii) the sum of subadditive functions, and (iii) $G(x) = \min\{f(x)+F(x-z): z \in [0,x]\}$ if f and F are subadditive.

2.5 If the optimum shipment size obtained with the construction of Fig. 5.8, v^*, is in the interval $((n-1) v'_{max}, nv'_{max})$, evaluate the difference $c(nv'_{max}) - c(v^*)$ and show that it cannot exceed $(5n)^{-1}(c_f/v'_{max})$.

2.6 Assume that a firm operating its own vehicle fleet uses the trigger point strategy of Section 2.5 and Fig. 2.9, with trigger point v_0 and shipment size v. A dispatched vehicle becomes available after a cycle time T_r that varies with every shipment; T_r can be viewed as a random variable (independent of all the others) with mean t_r and standard deviation σ_r. If the demand has the stochastic properties discussed in the text, prove that the fleet size needed to ensure that the firm does not run out of vehicles is of the form:

$$ D't_r/v + (\text{constant}) \left[(D'\sigma_r/v)^2 + (D'\gamma t_r/v^2) \right]^{1/2}, $$

where constant ≈ 3. (Hint: the fleet size should be as large as the maximum number of requests that are likely to be received during a vehicle cycle).

Comment: Multiplied by an appropriate constant, the second term of this expression is the contribution to transportation cost caused by uncertainty. Notice that it is proportional to $(1/v)$ and independent of v_0. Therefore, the optimal v is still given by the solution of Eq.

(2.8a) if the fixed transportation cost is duly modified. A modification to the transportation coefficient of Eq. (2.8a) can also be found if, in order to lessen the need for a larger fleet, we allow the inventory buffer at the destination to be increased in anticipation of (rare) vehicle shortages.

Glossary of Symbols

A:	EOQ formula constant, $A = c_h/D'$,
B:	EOQ formula constant, $B = c_f$,
c_d:	Transportation cost per vehicle-mile (\$/vehicle-distance),
c'_d:	Marginal transportation cost per item, per distance unit (\$/item-distance),
c_e:	Cost per item expedited, when using two shipping modes,
c_f:	Fixed transportation cost for a shipment, independent of size (\$/shipment),
c'_f:	Fixed handling cost of moving a pallet,
c''_f:	Fixed motion cost (transportation + handling) per shipment,
c_h:	Holding cost per item, per unit time, $c_h = c_r + c_i$,
c_i:	Waiting cost per item, per unit time (\$/item-time),
c_r:	Rent cost per item, per unit time (\$/item-time),
c_s:	Fixed transportation cost of stopping for a shipment (part of c_f independent of distance),
c'_s:	Added transportation cost of carrying an extra item (part of c_v independent of distance) (\$/item),
c_v:	Added transportation cost per extra item carried (\$/item),
c'_v:	Added handling cost per extra item handled,
c''_v:	Incremental motion cost per item moved (transportation + handling),
d:	Distance traveled,
D':	Demand rate (items/time),
D'_p:	Production rate (items/time),
$f_h()$:	Handling cost function per shipment,
$f_m()$:	Motion cost function per shipment, $f_m = f_t + f_h$,
$f_t()$:	Transportation cost function per shipment,
γ:	Index of dispersion of the demand arrival process (items),
H:	Generic headway between successive shipments (time), assumed to be the first in a sequence,
H_1:	Maximum interval between successive dispatches (time),
H_i:	Headway between the ith and the $(i+1)^{th}$ dispatch (time),
i:	Annual discount rate for money (\$/\$-year), used in Sec. 2.1 only,
n:	Number of shipments,
n_s:	Number of stops,
π:	Value of one item (\$/item),
π_0:	Production cost of one item,

π_1:   Selling price for one item,

$\overline{PQ}$:   Generic segment of a figure,

p_t: Costs incurred during year t,

σ_ℓ: Standard deviation of the lead time,

t: Time,

t_ℓ: Mean of the "lead time", T_ℓ,

T_ℓ :   Lead time (period between order placement and arrival),

t_m: Transportation time between origin and destination,

v: Generic shipment size (items),

V: Total number of items shipped,

$\overline{v}$:   Average shipment size,

v':   Average number of items shipped (regular plus expedited) per regular shipment,

v^*:   Optimal shipment size,

v_0:   Inventory trigger point,

v_i: Size of the ith shipment,

v_{max}:   Capacity of a vehicle (items),

v'_{max}:   Capacity of a pallet (items).

3 Optimization Methods: One-to-One Distribution

Readings for Chapter 3

Newell (1971) shows how to find an optimal sequence of headways for a transportation route serving a changing demand over time with a continuum approximation method that avoids "details." This problem is mathematically analogous to the problems with time dependent demand addressed in this chapter, which are traditionally solved with dynamic programming. Section 3.3, is based on this reference. Daganzo (1987) shows that a continuous approximation of a function and its variables can be *more* accurate than the exact, detailed and discontinuous world representation they replace. This result is discussed in Section 3.2.

3.1 Initial Remarks

This chapter describes logistics problems linking one origin and one destination (one-to-one problems) and the methods used to solve them. The following points, mentioned in Chapter 1, will be revisited:

(i) Accurate cost estimates can be obtained without precise, detailed input data,

(ii) Departures from an optimal decision by a moderate percentage do not increase cost significantly. Since there is no need to seek the most accurate estimate of the optimum, there may be little use for highly detailed data,

(iii) Detailed data may get in the way of the optimization, actually *hindering* the search for an optimum,

(iv) Thus, we advocate a two-step solution approach to logistics problems: the first (analytical) step involves little detail and yields broad solution concepts; the second (or fine tuning) step leads to specific solutions, consistent with the ideals revealed by the first – it uses all the relevant detailed information.

These points will be illustrated with simple extensions of the EOQ model introduced in Chapter 2. Section 3.2 analyzes one-to-one systems with constant production and consumption rates; the discussion focuses on the robustness and accuracy of the results. Section 3.3 examines the same problem when the demand varies over time; it describes numerical methods and a continuous approximation (CA) analytical approach that is based on summarized data. Section 3.4 illustrates how the CA approach can be used for a location problem that has an analogous structure, and Section 3.5 demonstrates the accuracy of the CA solutions.

As a prelude to the more complex problems explored in forthcoming chapters, Section 3.6 explains how the CA approach can be extended to multidimensional problems with constraints, and Section 3.7 discusses network design issues.

3.2 The Lot Size Problem with Constant Demand

Let us now explore the optimization problem for the optimum shipment size, v^*, described in the previous chapter:

$$z = \min\left\{Av + \frac{B}{v} : v \leq v_{max}\right\}. \tag{3.1}$$

Consider first the case $v_{max} = \infty$. Then v^* is the value of v which minimizes the convex expression $Av + Bv^{-1}$:

$$v^* = \sqrt{\frac{B}{A}}. \tag{3.2}$$

Remember that B represented the fixed motion costs, c_f, and A the holding cost per item, c_h/D'. Note that v^* is the value which makes both terms of the objective function equal. That is, for an optimal shipment size, holding cost = motion cost.

The optimum cost per item is:

$$z^* = (cost/item)^* = 2\sqrt{AB}, \tag{3.3}$$

which is easy to remember as "twice the square root of the product" of the two terms of (3.1). It will be convenient to memorize Eqs. (3.2) and (3.3), since EOQ minimization expressions will arise frequently.

As a function of c_f, c_h and D', the optimum cost per item increases at a decreasing rate with c_f and c_h and decreases with the item flow D'. There are *economies of scale*, since higher item flows lead to lesser average cost.

In the remainder of this section we examine the sensitivity of the resulting cost to errors in: (i) the decision variable, v , (ii) the inputs (A or B), and (iii) the functional form of the equation.

3.2.1   Robustness in the Decision Variable

Suppose that instead of v^* , the chosen shipment size is $v^0 = \gamma v^*$, where γ [1] is a number close to 1, capturing the relative error in v^0 . Then, the ratio of the actual to optimum cost z^0/z^* will be a number, γ' , greater than 1, satisfying:

$$\gamma' = \left[A \sqrt{\frac{B}{A}} \gamma + B \sqrt{\frac{A}{B}} \frac{1}{\gamma} \right] / \left[2 \sqrt{AB} \right] = \frac{1}{2} \left[\gamma + \frac{1}{\gamma} \right]. \qquad (3.4)$$

Independent of A and B , this relationship between input and output relative errors holds for all EOQ models. It indicates that if γ is between 0.5 and 2, so that the optimal shipment size is approximated to within a factor of 2, then $\gamma' \leq 1.25$. If γ is between 0.8 and 1.25, then $\gamma' \leq 1.025$. Thus, a cost within 2.5 percent of the optimum can be reached if the decision variable is within 25 percent of optimal. On the other hand, if γ is several times larger (or smaller) than 1, then the cost penalty is severe, i.e., $\gamma' \approx \gamma$ (or $\gamma' \approx 1/\gamma$) . Obviously, thus, while it is important to get reasonably close to the optimal value of the decision variable (say to within 20 to 40 percent), from a practical standpoint it may not be imperative to refine the decision beyond this level.

3.2.2   Robustness in Data Errors

Let us now assume that one of the cost coefficients A (or B) is not known precisely. If it is believed to be A' = Aδ (or B' = Bδ) , for some $\delta \approx 1$, then the optimal decision with this erroneous cost structure is:

---

[1] This symbol is unrelated to the coefficient of variation of the prior chapter, also denoted by γ.

$$v^{*'} = \left(\sqrt{\frac{B}{A}} \right) \delta^{-1/2} = v^* \, \delta^{-1/2} \qquad if \quad A' = A\delta \, ,$$

or

$$= v^* \, \delta^{1/2} \qquad\qquad if \quad B' = B\delta \, .$$

Because the actual to optimal shipment size ratio, $v^{*'}/v^*$, is either $\delta^{-1/2}$ or $\delta^{1/2}$ (see Eq. (3.2)), the cost penalty paid is as if $\gamma = \delta^{1/2}$. Thus, the resulting cost is even less sensitive to the data than it is to the decision variables. For example, if the input is known to within a factor of 2 $(0.5 \le \delta \le 2)$, then $0.7 \le \gamma \le 1.4$ and $\gamma' \le 1.1$. The cost penalty would be about 10 percent, whereas before it was 25 percent. The penalty declines quickly as δ approaches 1. This robustness to data errors is fortunate because, as we pointed out in Chapter 2, the cost coefficients (for waiting cost especially) are rarely known accurately.

3.2.3   Robustness in Model Errors

A cost penalty is also paid if the EOQ formula itself is inaccurate. To illustrate the impact of such functional errors, we assume that the actual cost, a complicated (perhaps unknown) expression, can be bounded by two EOQ expressions; the cost penalty can then be related to the width of the bounds.

Suppose, for example, that the actual holding cost $z_h(v)$ is not exactly equal to the EOQ term (Av), but it satisfies:

$$Av - \Delta/2 \le z_h(v) \le Av + \Delta/2$$

for some small Δ . (Such a situation could happen, for example, if storage space could only be obtained in discrete amounts.) Because Δ is small, the EOQ lot size v^* is adopted. Clearly, then the absolute difference between the actual cost $[z_h(v^*)+ B/v^*]$ and the predicted EOQ cost z^* cannot exceed $\Delta/2$. It is also easy to see that the difference between the optimal cost with perfect information, $\min\{z_h(v) + B/v\}$, and z^* cannot exceed $\Delta/2$ either. As a result, the difference between the actual and theoretical minimum costs – the cost penalty – is bounded by Δ.

Usually, though, this penalty will be significantly smaller than the maximum possible; Figure 3.1 illustrates the unusual conditions generating the largest penalty. Thus, if Δ is small compared to z^* (e.g., within 10 percent) the functional form error should be inconsequential. The same conclusion is reached if the motion cost is also inaccurate. In general, the EOQ solution will be reasonable if it is accurate to within a small fraction of its predicted optimal cost.

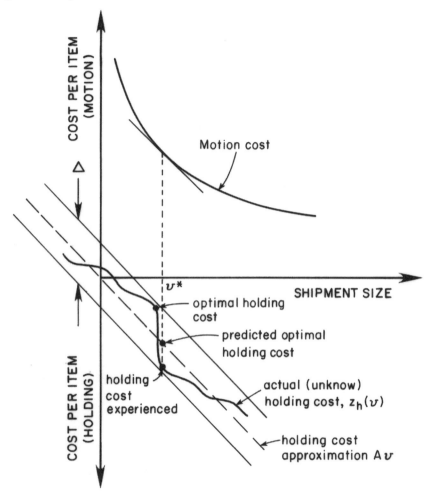

Fig. 3.1   Cost penalty resulting from errors in the holding cost function

3.2.4 Error Combinations

If errors of the three types exist, one would expect the cost penalty to be greater. Fortunately though, when dealing with errors the whole (the combined penalty) is not as great as the sum of its parts.

Suppose for example that the lot size recipe is not followed very precisely (because, e.g., lots are chosen to be multiples of a box, only certain dispatching times are feasible, etc.) and that as a result 40 percent discrepancies are expected between the calculated and actual lot sizes. We have already seen that such discrepancies can be expected to increase cost by about 10 percent. Let us assume, in addition, that one of the inputs (A or B) is suspected to be in error by a factor of 2, which taken alone would also increase cost by about 10 percent. Would it then be reasonable to expect a 20 percent cost increase? The answer is no; it should be intuitive that the penalty paid by introducing an input error when the lot size decision does not follow the recipe accurately should be smaller than the penalty paid if the decision follows the recipe. In our example, the combined likely increase is 14 percent [the square root of the sum of the squared errors: $.14 = (.1^2 + .1^2)^{1/2}$]. Statistical analysis of error propagation through models reveals similar composition laws in more general contexts (see e.g., Daganzo, 1985). This subject, however, is beyond the scope of this monograph. Further information can be found in Taylor (1997).

The above example illustrated how input and decision errors propagate. Although model errors follow similar laws – the whole is still less than the sum of the parts – for some approximate models the results are surprising. The composed (data and model) error can be actually *smaller* than the data error alone with the exact model! (Daganzo, 1987). This fortuitous phenomenon, illustrated by problem 3.1, has a special significance because it arises when, as recommended in this monograph, certain discontinuous models with discrete inputs are approximated by continuous functions and data. A more detailed discussion of this issue can be found in Daganzo (1987).

For ease of exposition, our discussion of robustness and errors ignored the $v \leq v_{max}$ constraint of Eq. (3.1), although similar remarks could have been made for the constrained solution and other non-EOQ models (see exercise 3.10). The constrained EOQ solution is now presented rather briefly, before turning our attention to the lot size problem with variable demand.

If, in solving the unconstrained EOQ problem, we find that $v^* > v_{max}$, then the solution is not feasible. In that case, choosing $v = v_{max}$ is optimal. Hence, the optimal EOQ solution can be expressed as:

$$v^* = \min\left\{\sqrt{\frac{B}{A}}, v_{max}\right\}, \tag{3.5a}$$

and the optimal cost per item z^* is:

$$
\begin{aligned}
z^* &= 2\sqrt{AB} && \text{if } \sqrt{B/A} \le v_{max} \\
&= Av_{max} + \frac{B}{v_{max}} && \text{if } \sqrt{B/A} > v_{max}
\end{aligned}
\tag{3.5b}
$$

Note that z^* is an increasing and concave function of A , and also of B (see Fig. 3.2a and b). As a function of $1/A = D'/c_h$, and thus of D' , z^* is decreasing and convex; the economies of scale continue to exist for all ranges of D'. Finally, note that the total cost per unit time, $D'z^*$, is proportional to $D'^{1/2}$ until the capacity constraint is reached, and from then on increases linearly with D'. The critical point is $D'_{crit} = (v_{max})^2/c_f$. The general form of the relationship is depicted in Fig. 3.2c.

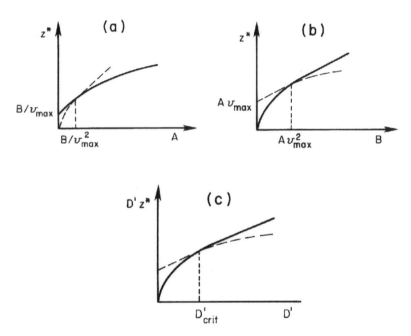

Fig. 3.2   Optimal EOQ cost as a function various parameters: (a) holding cost per item, A; (b) fixed motion costs, B; and (c) demand rate, D'. Dashed lines are the unused branches of Eq. (3.5b)

3.3 The Lot Size Problem with Variable Demand

Let us now consider the EOQ problem over a finite time horizon when the consumption rate D' changes with time in a predictable manner. The demand pattern, an input to our problem, is characterized by a function $D(t)$ that gives the cumulative number of items demanded between times 0 (the beginning of the study period) and t . The time derivative of this function $D'(t)$ represents the variable demand rate. We then seek the set of times when shipments are to be received ($t_0 = 0$, t_1 , ... , t_{n-1}), and the shipment sizes (v_0, v_1 , ... , v_{n-1}), that will minimize the sum of the motion plus holding costs over our horizon, $t \in [0, t_{max}]$.

As in Chapter 2, we also define as inputs to our problem a fixed (motion) cost per vehicle dispatch c_f , a holding cost per item-time $c_h = c_r + c_i$, and a maximum lot size v_{max} . With an infinite horizon and a constant demand, $D(t) = D't$, this formulation reduces to the EOQ problem examined in Section 3.2, where $A = c_h/D'$ and $B = c_f$.

For most of this section, we assume that the v_{max} constraint can be ignored. We will relax this restriction in Section 3.6. Subsection 3.3.1, below, examines the variable demand problem when rent costs are the dominant part of holding cost; a simple solution can then be obtained. Subsection 3.3.2 shows that if inventory (waiting) costs are dominant, then the solution is not quite as apparent; two solution methods are then described: a numerical method in subsection 3.3.3 and an analytical method in subsection 3.3.4.

3.3.1 Solution When Holding Cost Is Close to the Rent Cost

If inventory cost is negligible, $c_i \ll c_r$, then holding cost approximately equals rent cost $c_h \approx c_r$. We have already mentioned that rent cost increases with the *maximum* inventory accumulation (regardless of when it is held), and that otherwise the cost is rather insensitive to the accumulations at other times. This property of holding cost simplifies the solution to our problem.

Recall from Sec. 2.3 that given a set of n shipments, the motion cost during the period of analysis, $c_f n$, is independent of the shipment times and sizes. The problem, then, is to find the sets of shipment times and sizes that will minimize holding cost. A lower bound to the maximum accumulation at the destination is the size of the largest shipment received, which is minimized when all the shipments are equal. Hence, the largest shipment – and, thus, the maximum accumulation – must exceed or at least equal $D(t_{max})/n$. If a set of times and shipment sizes is found for which the maximum accumulation equals $D(t_{max})/n$, the set is an optimal way of sending n shipments with rent cost per unit time: $c_r D(t_{max})/n$. Figure 3.3 depicts such a solution for a hypothetical cumulative consumption curve $D(t)$. Each shipment is just large enough to meet the demand until the next shipment; the consumption between consecutive receiving times, the same in all cases, is $D(t_{max})/n$. Clearly then, the following strategy is optimal:

(i)   Divide the ordinate axis between 0 and $D(t_{max})$ into n equal segments and find the times t_i for which $D(t)$ equals $(i/n)D(t_{max})$ for $i = 0, ... , n - 1$. These are the shipment times,

(ii)   Dispatch barely enough to cover the demand until the following shipment.

One must now find the optimal n by minimizing the resulting cost. Interestingly, it does not depend on the t_i , only on n:

$$cost/time = c_r \left(D(t_{max})/n \right) + c_f \left(n/t_{max} \right), \quad and$$

$$cost/item = \left(\frac{c_r}{\overline{D'}} \right) \left(\frac{D(t_{max})}{n} \right) + c_f \left(\frac{n}{D(t_{max})} \right) , \qquad (3.6)$$

where $\overline{D'}$ is the average consumption rate:

$$\overline{D'} = D(t_{max})/t_{max} .$$

Note that (3.6) is the EOQ expression with $v = D(t_{max})/n$. The solution now requires that n be an integer (there are constraints on v), but we have already seen that any v close to the unconstrained v* is near optimal. As a result, unless the time horizon is so short that n* = 1 or 2, the optimal cost per item should be close to the cost with constant demand.

It should be intuitive that if $v_{max} < \infty$, the solution procedure does not change. It is still optimal to have equal shipment sizes, but the number of shipments should be large enough to satisfy: $D(t_{max})/n \leq v_{max}$. The solution is still of the form (3.5), with v^{-1} restricted to being an integer multiple of $D(t_{max})^{-1}$.

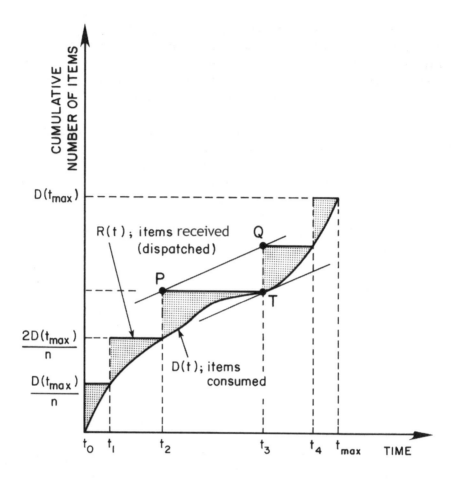

Fig. 3.3   Selection of shipment times for least holding cost

3.3.2 *Solution when Rent Cost Is Negligible*

Let us now examine another extreme but common situation, where items are so small and expensive, that most of the holding cost arises from the item-hours spent in inventory, and not from the rent for the space to hold them. In this case the destination's holding cost should be proportional to the shaded area of Fig. 3.3.

The combined origin-destination holding cost will also be proportional to this area if (i) the origin holding cost can be ignored, or (ii) if it is proportional to the area. Situation (i) arises if the origin produces generic items for so many destinations that the part of its costs that would be prorated to each destination is negligible. The second situation arises if the production strategy at the origin is as described in Fig. 2.10. Then, we see from that figure that the total wait at the origin that can be attributed to the shipping strategy must be similar to that of the destination; i.e., it would also be proportional to the shaded area of Fig. 3.3. A third scenario arises with typical passenger transportation systems.

When holding costs are proportional to the area of Fig. 3.3 they are no longer a function of n alone. Newell (1971) points out that for a set of points $(t_1 \ldots t_{n-1})$ to be optimal, each line $\overline{PQ}$ (of Fig. 3.3) must be parallel to the tangent line to $D(t)$ at the receiving time (point T in the figure). The reader can verify that if this condition is not satisfied, then it is possible to reduce the total shaded area by either advancing or delaying the receiving time by a small amount.

Unfortunately, the smallest shaded area – and thus the waiting cost – no longer can be expressed as a function of n alone, independently of $D(t)$. Thus, it seems that a simple expression for the optimal cost cannot be obtained for any $D(t)$. (Subsection 3.3.4 develops an approximation when $D(t)$ varies slowly with t).

3.3.3 Numerical Solution

There are different ways in which this problem can be solved numerically. For example, it can be formulated as a dynamic program in which a shipment time, t_i , is chosen at each stage $(i = 1 , \ldots, n - 1)$, and where the state of the system is the prior shipment time, t_{i-1} . The dynamic programming procedure yields an optimum holding cost for a given n , $z^*_i(n)$, which can be substituted for the first term of Eq. (3.6) to yield n^*.

The following procedure, based on Newell's property, is less laborious and works particularly well if $D(t)$ is smooth, without bends or jumps (refer to Figure 3.4 for the explanation):

(i)   Choose a point P_1 on the ordinates axis and move across to T_1,
(ii) Draw from P_1 a line parallel to the tangent to $D(t)$ at T_1, and draw from T_1 a vertical line. Label the point of intersection P_2 .

Steps (i) and (ii) identify a point P_2 from a point P_1 . They should be repeated to identify P_3 from P_2, P_4 from P_3, etc. ..., defining in this manner a receiving step curve, $R(t)$. If $R(t)$ does not pass through the end point, $(t_{max} , D(t_{max}))$, the position of P_1 should be perturbed until it does.

If a different point P_1 is chosen, a different number of steps may result, and the motion cost will change.[2] The holding cost for the given P_1 is proportional to the area between R(t) and D(t) ; it will also change if P_1 is moved. The overall optimum can be found by shifting the position of P_1 and comparing the sum of the holding and motion costs.

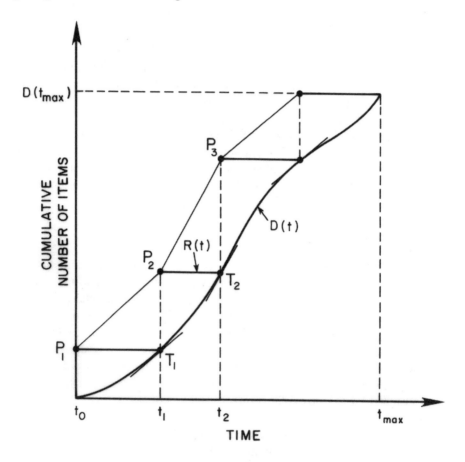

Fig. 3.4   Construction method for the cumulative number of items shipped versus time

3.3.4 The Continuous Approximation Method

The method about to be described, proposed by Newell (1971), replaces the search for $\{t_i\}$ by a search for a continuous function, whose knowledge

---

[2] As illustrated with problem 3.6, there may be more than one solution with the same number of steps.

yields a set of t_i with near minimal cost. It works well when $D'(t)$ does not change rapidly; i.e., if $D'(t_i) \approx D'(t_{i+1})$ for all i. A by-product is a simple expression and decomposition principle for the total cost.

Let us assume that an optimal solution has been found, and denote by I_i the ith interval between consecutive receiving times: $[t_{i-1}, t_i)$, $i = 1, 2, ...$. Then, divide the total cost during the study period into portions "$cost_i$" corresponding to each interval. That is, "$cost_i$" includes the cost, c_f, of dispatching one shipment plus the product of c_i and the shaded area for interval I_i:

$$cost_i = c_f + c_i(area_i).$$

Clearly, the sum of the prorated costs will equal the total cost. Since $D'(t)$ is continuous, it should be intuitive that there is a point t'_i in each interval I_i for which the area above $D(t)$ satisfies: $area_i = \frac{1}{2}(t_i-t_{i-1})^2 D'(t'_i)$. To see this informally, consider the triangle defined by the horizontal and vertical lines passing through a point P_i in the figure and a straight line passing through T_i with a slope that yields "$area_i$" for the triangle; i.e. slope $D'(t'_i)$. Since such a slanted line must intersect $D(t)$ (otherwise the areas above $D(t)$ and above the slanted line could not be equal) there must be a point between T_i and the point of intersection where the two lines have the same slope. The abscissa of this point is t'_i. Therefore we can write:

$$area_i = \frac{1}{2}\left(t_i - t_{i-1}\right)^2 D'\left(t'_i\right) = \int_{t_{i-1}}^{t_i} \frac{1}{2}\left(t_i - t_{i-1}\right)D'\left(t'_i\right)dt. \tag{3.7}$$

If we now define $H_s(t)$ as a step function such that $H_s(t) = t_i - t_{i-1}$ if $t \in I_i$ (see Figure 3.5 for an example), then the cost per interval can be expressed as:

$$cost_i = \int_{t_{i-1}}^{t_i}\left[\frac{c_f}{H_s(t)} + \frac{c_i H_s(t)}{2}D'\left(t'_i\right)\right]dt. \tag{3.8}$$

Note that this is an exact expression.

If we now approximate $D'(t'_i)$ by $D'(t)$ – which is reasonable if $D'(t)$ varies slowly – the total cost over the whole study period can be expressed as the following integral:

$$cost \approx \int_{0}^{t_{max}}\left[\frac{c_f}{H_s(t)} + \frac{c_i H_s(t)}{2}D'(t)\right]dt. \tag{3.9}$$

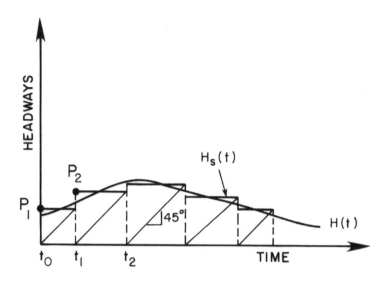

Fig. 3.5   Obtaining a set of dispatching times from H(t)

We seek the function $H_s(t)$, which minimizes (3.9). Unfortunately, this is akin to determining the $\{t_i\}$ themselves. A closed form solution can be obtained if in (3.9) $H_s(t)$ is replaced by a smooth function, $H(t)$, as shown in Fig. 3.5. That is:

$$cost \approx \int_0^{t_{max}} \left[\frac{c_f}{H(t)} + \frac{c_i H(t)}{2} D'(t) \right] dt . \qquad (3.10)$$

Now, instead of finding $H_s(t)$, we can find the $H(t)$ which minimizes (3.10) – a much easier task – and then choose a set of shipment times (i.e., $H_s(t)$) consistent with $H(t)$.

   Clearly, the $H(t)$ which minimizes (3.10) minimizes the integrand at every t; thus:

$$H(t) = \left[2c_f / \left(c_i D'(t) \right) \right]^{1/2} . \qquad (3.11a)$$

This is the time between dispatches (*headway*) for the EOQ problem with constant demand $D' = D'(t)$.

A set of shipment times consistent with H(t) can be found easily since H(t) varies slowly with t ; see (3.11a). Figure 3.5 suggests how this can be done systematically: Starting at the origin (point t_0) draw a 45° line and find a horizontal segment from a point on the vertical axis, such as P_1 in the figure, to the intersection with the 45° line. The elevation of P_1 should be such that the area below the segment equals the area below H(t). The abscissa of the point of intersection is the next shipment time, t_1 . This locates t_1 , given t_0 . The construction is then repeated from t_1 to locate t_2 , from t_2 , to locate t_3 , etc. In practice one does not need to be quite so precise, since we have already seen that small deviations from optimality have a minor effect.

Replacing the right side of (3.11a) for H(t) in integral (3.10) yields a simple expression for the optimal cost:

$$Total\ cost \approx \int_0^{t_{max}} \left[2c_i c_f D'(t) \right]^{1/2} dt . \qquad (3.11b)$$

The integrand of this expression is the optimal EOQ cost per unit time if D' = D'(t).

Note that the integrand of Eq. (3.11b) can be written as:

$$\left[2c_i c_f / D'(t) \right]^{1/2} \left[D'(t) dt \right],$$

where the first factor represents the optimal cost *per item* for an EOQ problem with constant demand, D'(t) ; see Eq. (3.3). The average cost per item (across all the items) is obtained by dividing (3.11b) by the total number of items,

$$D(t_{max}) = \int_0^{t_{max}} D'(t) dt$$

The result is:

$$\left(\frac{cost^*}{item} \right) \approx \frac{\int_0^{t_{max}} \left[2c_i c_f / D'(t) \right]^{1/2} D'(t) dt}{\int_0^{t_{max}} D'(t) dt} . \qquad (3.11c)$$

In practical terms this equation indicates that the average optimal cost per item can be obtained by *averaging the cost of all the items, as if each one of these was given by the EOQ formula with a (constant) demand rate equal to the demand rate at the time when the item is consumed.*

Equation (3.11b) has a similar (decomposition) interpretation: the expression indicates that, given a partition of $[0, t_{max}]$ into a collection of short time intervals, the optimum cost can be approximated by the sum of the EOQ costs for each one of the intervals considered isolated from the others.

Equations (3.11) are so simple that they can be used as building blocks for the study of more complex problems as we shall see in later chapters. This is one of the attractive features of the CA approach; it yields cost estimates without having to develop, or even define, a detailed solution to the problem.

The CA approach can also be used to locate points on any line (time or otherwise) *provided that the total cost can be prorated approximately to (short) intervals on the line, while ensuring that the prorated cost to any interval only depends on the characteristics of said interval.* In the previous discussion, the integrand of (3.10) is the prorated cost in $[t, t+dt)$, which does not depend on the demand rate outside the interval.

The CA approach can also be used to locate points in multidimensional space, when the total cost can be expressed as a sum of neighborhood costs dependent only on their local characteristics. Newell (1973) argues that the CA approach is comparatively more useful then, because in the multidimensional case it is much more difficult for exact numerical methods to deal with the complex boundary conditions that arise. Because the CA approach will be used in forthcoming chapters repeatedly, the next section discusses two additional (one-dimensional) examples.

3.4   Other One-Dimensional Location Problems

The CA technique was originally proposed to find a near-optimal bus departure schedule from a depot (Newell, 1971). Given the cumulative number of people $D(t)$ demanding service by time t, the fixed cost of a bus dispatch c_f, and the cost of each person-hour waited c_i, the objective was to minimize the sum of the bus dispatch (motion) and waiting (holding) costs. With an unlimited bus capacity, this problem is almost identical to the one we have just solved; except for $D(t)$, which now represents the cumulative number of people (items) *entering* the system and not the number leaving. Equations (3.11), however, still hold (see problem 3.2) This should be intuitive. Although the graphical construction of Figure 3.4 is now slightly different (i.e., the sought passenger departure curve R(t) now touches D(t) from below) consideration shows that the new and old figures become qualitatively identical if one of them is rotated 180 degrees. Since such a rotation cannot change the mathematical relationships between the elements of the figure, it shouldn't be surprising that Eqs. (3.11) remain valid.

The second example locates freight terminals on a distance line between 0 and d_{max}. This interval contains origins, which send items to a depot.

The distance line extends from the origin, O , to a depot, located at d = $\tilde{d} \geq d_{max}$. The flow of freight (number of items per day) that originates between O and d is a function of d , D(d), which increases from 0 to v_{tot} (see Figure 3.6). Items are individually carried to the terminals at a cost c_d' per unit distance per item. Each day a vehicle travels the route collecting the items accumulated at each terminal and takes them to the depot.

The motion cost for this operation has three components: the handling cost at the terminals, assumed to be constant and therefore ignored, the access cost to the terminals, and the line-haul cost of operating the vehicle from the terminals to the depot. The access cost is given by the product of c_d' and the total item-miles of access traveled per day; it increases with the separation between stops as will be explained in a moment. The line-haul cost has the form of Eq. (2.5d):

$$\left(\frac{line\text{-}haul}{cost/day}\right) = c_s\left(1+n_s\right)+c_d\left(\tilde{d}\right)+c'_s\left(v_{tot}\right),$$

where n_s is the number of stops (excluding the depot) and v_{tot} is the total size of the shipment arriving at the depot. Note that the line-haul cost does not depend on the specific stop locations and that in contrast to the access cost, it increases with n_s. As a function of n_s we express it as:

$$\left(\frac{line\text{-}haul}{cost/day}\right) = c^o + c_s n_s , \qquad (3.12)$$

where c^o is a constant that will be ignored for design purposes.

As the problem has been formulated, with one trip per day, the sum of the holding costs at all stops can be ignored – consideration reveals that the sum is constant. Pipeline inventory costs do depend on the decision variables (they should increase with n_s) but for cheap freight the effect is negligible relative to (3.12). Thus, all inventory and holding costs are neglected. The stops will be located as the result of a trade-off between line-haul and access costs. Without this simplification, which is inappropriate for passenger transportation, the problem is equivalent to the transit stop location problem solved by Vuchic and Newell (1968) with dynamic programming, and later by Hurdle (1973), and Wirasinghe and Ghoneim (1981) with the CA method. (See problem 3.3).

Figure 3.6 depicts the location of three terminals (at points d_1, d_2, and d_3) and a curve, $R(d)$, depicting the number of items in the vehicle as a function of its position. This curve increases in steps at each terminal location. The size of each step equals the number of items collected. To minimize access (and total) cost each item is routed to the nearest terminal, and as a result the step curve passes through the midpoints, M_i, shown in the figure. (The coordinates of M_i are $m_i = (d_i + d_{i+1})/2$ and $D(m_i)$; with $m_0 = 0$ and $m_{ns} = d_{max}$).

Let us see how the total cost can be prorated to short intervals, by considering the partition of $(0, d_{max}]$ into the following intervals surrounding each terminal: $I_1 = (0, m_1]$, $I_2 = (m_1, m_2]$, ... , $I_{ns} = (m_{ns-1}; d_{max}]$. Each interval, I_i, adds an access cost proportional to the daily item-miles traveled for access to terminal i. This is given by the shaded area on the two quasi-triangular segments next to the location of the terminal, $(area)_i$, thus:

$$access\ cost_i = (area)_i\ c'_d .$$

For slowly varying $D(d)$, the access cost can be rewritten as:

$$access\ cost_i \approx \frac{1}{4}(m_i - m_{i-1})^2\ D'(d_i)c'_d .$$

Since each terminal adds c_s to the daily line-haul cost (see Eq. (3.12)), the share of the total cost prorated to I_i is:

$$\left(\frac{Total\ cost}{per\ day}\right)_i \approx c_s + \frac{c'_d}{4}(m_i - m_{i-1})^2\ D'(d_i).$$

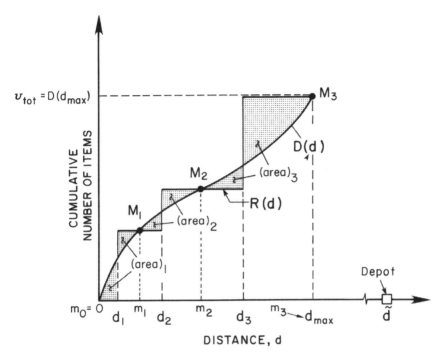

Fig. 3.6   Geometrical construction for a terminal location problem

Since $D'(d) \approx D'(d_i)$ for $d \in I_i$ (we stated that $D'(d)$ varied slowly), the above expression can be approximated by:

$$\begin{pmatrix} Total\ cost \\ per\ day \end{pmatrix}_i \approx \int_{m_{i-1}}^{m_i} \left\{ \frac{c_s}{(m_i - m_{i-1})} + \frac{c'_d}{4}(m_i - m_{i-1})D'(d) \right\} dd\ .$$

If we now let $s(d)$ denote a slowly varying function such that $s(d_i) = m_i - m_{i-1}$ (the function, used later to locate the terminals, indicates the size of a terminal's influence area depending on location), then we can re-write the last expression once again, using $s(d)$ instead of $m_i - m_{i-1}$:

$$\begin{pmatrix} Total\ cost \\ per\ day \end{pmatrix}_i \approx \int_{m_{i-1}}^{m_i} \left\{ \frac{c_s}{s(d)} + \frac{c'_d}{4}s(d)D'(d) \right\} dd\ .$$

The total cost for the system is then:

$$\left(\begin{array}{c}\text{Total cost}\\ \text{per day}\end{array}\right) \approx \int\limits_{0}^{d_{max}} \left\{ \frac{c_s}{s(d)} + \frac{c'_d}{4} s(d) \, D'(d) \right\} dd . \tag{3.13}$$

As with Eqs. (3.11), the least cost $s(d)$ minimizes the integrand at every point; given its EOQ analytical form, we find:

$$s(d) \approx 2\left[c_s \, / \, \left(c'_d \, D'(d) \right) \right]^{1/2} . \tag{3.14a}$$

(Note that if D' varies slowly, $s(d)$ will vary slowly as we had assumed.)

The expressions for the minimum total and average (per item) cost are similar to (3.11b) and (3.11c); the partition/decomposition principle still holds.

$$\left(\begin{array}{c}\text{Total cost}\\ \text{per unit}\\ \text{of time}\end{array}\right)^{*} \approx \int\limits_{0}^{d_{max}} \left[c_s c'_d \, D'(d) \right]^{1/2} dd , \tag{3.14b}$$

$$Cost^* /item \approx \int\limits_{0}^{d_{max}} \left[c_s c'_d / D'(d) \right]^{1/2} D'(d) \, dd \, / \int\limits_{0}^{d_{max}} D'(d) \, dd . \tag{3.14c}$$

To locate the terminals, one first divides $(0, d_{max}]$ into non-overlapping intervals of approximately correct, length I_1, I_2, etc. ..., by starting at one end and using (3.14a) repeatedly. If the last interval is not of correct length, then the difference can be absorbed by small changes to the other intervals. If d_{max} is large (so that there are at least several intervals), then the final partition should satisfy $s(d) \approx m_i - m_{i-1}$ if $d \in I_i$, and the approximations leading to (3.14) should be valid. With the influence areas defined in this manner, the terminals are located next. They should be positioned within each interval so that the boundary between neighboring intervals is equidistant from the terminals. For a general sequence of intervals (e.g., of rapidly fluctuating lengths) this may be difficult (even impossible) to do, but for our problem with $|I_i| \approx |I_{i+1}|$ the best locations should be near the center of each interval; in fact little is lost by locating the terminals at the centers.

3.5   Accuracy of the CA Expression

Although a systematic analysis of its errors has not been reported, experience indicates that the CA approach is very accurate when the descriptive characteristics of the problem (D'(t) in the text's examples) vary slowly as assumed. Also quite robust, the approach is effective even if the variation in conditions is fairly rapid – in our case, accurate results are obtained even if D'(t) varies by a factor of two within the influence areas. Perhaps this should not be surprising, in light of the EOQ robustness discussed in Section 3.2.

When conditions are unfavorable, the CA method can both over- and underpredict the optimal cost. The following two examples identify said conditions, with the first example illustrating over-estimation and the second underestimation. The basis for comparison will be the exact solution, which for our problem can be obtained readily, as described below.

3.5.1   An Exact Procedure and Two Examples

A construction similar to that in Fig. 3.4 can also be used for the terminal location problem.

Note first that, given n_s, for a set of locations to be optimal the line $D(d)$ of Fig. 3.6 must bisect in two equal halves every vertical segment of $R(d)$. Otherwise, the terminal (e.g., terminal 3 of Fig. 3.6) could be moved slightly to decrease access cost. The optimal solution can then be found by comparing all the possible $R(t)$ with the above property.

For a given d_1, draw a vertical step that is bisected by $D(d)$, and move across horizontally so that the horizontal segment is also bisected by $D(t)$. This identifies d_2. Repeat the construction to find d_3, d_4, etc. (Only those values of d_1 for which the last vertical segment is bisected by $D(t)$ need to be considered seriously.) The optimal solution corresponds to a d_1 which minimizes the sum of the stop cost and access cost. The procedure is so simple that it can be implemented in spreadsheet form. (The user selects d_1 and the spreadsheet returns the graphs, and the cost; it is then easy to find the solution either interactively or automatically with the computer.) The examples can now be discussed.

Example 1:

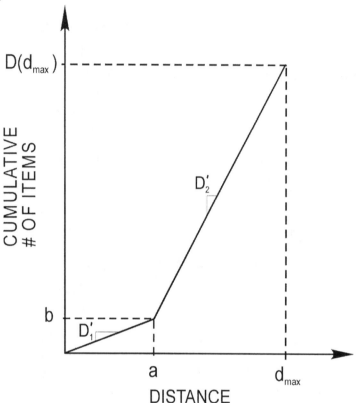

Fig. 3.7   Cumulative demand versus distance for example 1

Terminals are to be located on two adjoining regions with high and low demand. Figure 3.7 depicts a generic piece-wise linear cumulative demand curve of this type. The coordinates of the break-point (distance, item number) are given by parameters "a" and "b". They, of course must be consistent with the specified values for d_{max}, $D(d_{max})$, D'_1 and D'_2. For this problem the continuum approximation approach yields – see Eq. (3.14b):

$$Total\ cost^* \approx (c_s c'_d)^{1/2} \left\{ a\sqrt{D'_1} + (d_{max} - a)\sqrt{D'_2} \right\}$$

A possible set of parameters is $d_{max} = 500$, $D(d_{max}) = 1700$, $D'_1 = 1$, $D'_2 = 5$, $a = b = 200$, $c'_d = 1$ and $c_s = 160,000$. This choice has been made because a systematic analysis shows that it produces the largest overprediction error in percentage terms. The predicted cost is:

$$Total\ cost * \approx 348,328.$$

In actuality the least possible cost is 8% smaller. It arises when a single terminal is located at $d = 330$. The reader can verify that the exact access cost for this location is 160,500 units. Since the terminal cost is 160,000 units (for one terminal), the grand total is 320,500 < 348,328.

This rather extreme example illustrates that the CA approach can overestimate the optimum cost. To understand why this happens let us decompose the CA costs into its components. Note first that the ideal spacing between terminals predicted by the CA method with (3.14a) is:

$s(d) = 800$ in the low demand section, and
$s(d) = 357$ in the high demand section.

Thus, the CA access cost is calculated as if the average access distance was $s(d)/4 = 200$ in the low demand section and 89.25 units in the high demand section. Since there are 200 items in the low density region and 1500 in the high density region, the total CA access cost is approximately: $200 \times 200 + 89.25 \times 1500 \approx 173,875$. The CA stop cost is calculated by integrating the density of terminals over the service region, $(200/800 + 300/357) \approx 1.09$, and multiplying this result by the cost of a terminal: $1.09 \times 160,000 = 174,400$. The grand total is therefore: $173,875 + 174,400 = 348,275 \approx 348,328$.

It turns out, however, that just a single terminal in the high density region can serve both, the low density points with an average distance barely greater than the CA access distance, and the high demand section with an average access distance considerably inferior to the corresponding CA distance. For our chosen location ($d = 330$) the actual average access distances are: 230 units for the low density section (200 with the CA method) and 76 for the high density section (89 with CA method). Since we are using only one terminal, the final cost is lower.

The overprediction effect arises because the demand curve varies significantly and very favorably between the terminal and the edge of the service region, and the CA approach does not exploit this variation. The variation is so favorable that it allows a terminal provided for the high density points to double up efficiently as a terminal for the low density points.

Favorable conditions are unusual, however. When the demand does not vary rapidly the CA approach consistently underestimates demand.

Example 2: An example where the CA approach underestimates cost is easy to construct. By its nature, the CA approach ignores that the number of terminals must be an integer; any situation with a finite region size (or time horizon) will exhibit this error type. To exclude the overprediction error type illustrated by example 1, the demand per unit length of region is set constant: $D'(d) = D'$. This also allows closed form comparisons to be made.

The CA solution (3.14b) is:

$$Total\ cost^* = \sqrt{c_s c'_d}\ \sqrt{D' d_{max}}$$

Without losing generality, we choose the units of distance, item quantity and money so that $d_{max} = 1$, $D(d_{max}) = 1$ and $c_s = 1$. Thus, $D' = 1$ and only the parameter c'_d remains. The above expression becomes:

$$Total\ cost^* = \sqrt{c'_d}\ . \tag{3.15a}$$

If the exact optimal solution has n_s terminals, the distance line will be partitioned into n_s intervals of equal length: $I_i = ((i - 1)/n_s, i/n_s]$. The total cost is then:

$$Total\ cost\ (n_s) = n_s + 2n_s \left\{ \left(\frac{1}{2\,n_s} \right)^2 \frac{c'_d}{2} \right\} = n_s + \frac{c'_d}{4n_s} \tag{3.15b}$$

which is an EOQ expression in n_s . Its minimum over $n_s = 1, 2, 3, \dots$ is the optimal cost.

This least cost will always be greater or equal to the right side of Eq. (3.15a) because (3.15a) is the minimum of (3.15b) with unrestricted n_s , obtained for $n^*_s = (c'_d/4)^{1/2}$. Clearly, the underprediction will be most significant when n^*_s is close to an odd multiple of 0.5, or close to zero. Equation (3.4), which described the sensitivity of the EOQ cost expression to errors in the decision variables, also quantifies this underprediction; as n^*_s increases the underprediction quickly vanishes. Once $c'_d > 16$ (n^*_s is greater than 2) the difference is below one percent. If $c'_d > 4$ (the value at which $n^*_s = 1$) then the maximum difference stays below six percent. Although for smaller c'_d the difference can grow arbitrarily large as $c'_d \rightarrow 0$, that is not the case that is likely to be of interest; the large spacing between terminals recommended by the CA method (much larger than d_{max}) indicates that operating line-haul vehicles is probably an overkill. If it were of interest, and a terminal had to be provided, one could force the solution to the CA approach to satisfy the constraint $n_s \geq 1$. The next section will discuss how more involved constraints can be accommodated within a general CA framework.

Although exhibiting different errors types, both examples shared a common trait when their errors were largest: the ideal terminal spacing in an interval with constant demand exceeded the length of the interval; i.e., demand varied significantly within the spacing. Errors arose because this property violates the stated requirement for the CA approach: $D'(d)$ should vary slowly over distances comparable with $s(d)$. Conversely, the numerical results prove that an error below one percent results if $D(d)$ is piecewise linear with segments at least three times as long as each $s(d)$. Thus, any demand function that can be approximated in this manner should also yield accurate results.

3.6   Generalization of the CA Approach

The CA method can be applied to more complex problems – even problems that defy exact numerical solution. In forthcoming chapters it will be used to locate points in multidimensional (time-space) domains while satisfying decision variable constraints.

All that is needed is that the input data vary slowly with position, either in one or multiple dimensions, that the total cost can be expressed as a sum of costs over non-overlapping (small) regions of the location domain, and that these component costs (and constraints) depend only on the decisions made in their regions. If this is true, the decomposition principle holds and the CA results approximate the optimal cost accurately.

As a one-dimensional illustration, let us return to the inventory control problem of Eqs. (3.7) to (3.11), and let us assume that there is a capacity constraint on shipment size:

$$D(t_i) - D(t_{i-1}) \leq v_{max} .$$

This constraint has a local nature because it only involves quantities determined by events close to the time of shipment; i.e., by two neighboring dispatching times and by the amount of consumption between them. For any time t, thus, it should be possible to write the constraint approximately as an inequality including only variables and data specific to time t.

Recalling the definition of $H_s(t)$ (see Fig. 3.6), and using the slow-varying property of $D'(t)$, we can write:

$$D(t_i) - D(t_{i-1}) \approx H_s(t)D'(t) \approx H(t)D'(t)$$

and the constraint can be replaced by the approximation *based only on conditions at t:*

$$H(t)D'(t) \leq v_{max}, \quad or \quad H(t) \leq v_{max}/D'(t),$$

which must be satisfied for all t.

An approximate solution to our problem, thus, is an $H(t)$ that minimizes (3.10) subject to this constraint. The solution is of the form indicated by Eqs. (3.5); i.e., the optimal $H(t)$ is the least of: (i) the right side of (3.11a), $(2c_f/c_i\, D'(t))^{1/2}$, and (ii) $v_{max}/D'(t)$. Letting $\Psi\{x\}$ denote the increasing concave function $\{x^{1/2}$ if $x \leq 1$; or $\frac{1}{2}[1 + x]$ if $x > 1\}$, we can express the minimum cost per unit time concisely in terms of the dimensionless quantity, $2c_fD'(t)/(c_i\, v_{max}^2)$:

$$c_i v_{max}\psi\left\{2c_f D'(t)/c_i v_{max}^2\right\}.$$

Integrated from 0 to t_{max} , this expression approximates the optimal total cost, as in Eq. (3.11b). Note that when the argument of Ψ is less than one, as would happen if v_{max} is very large, then the expression coincides with the integrand of (3.11b), $[2c_ic_fD'(t)]^{1/2}$. An average cost per item can also be obtained as in Eq. (3.11c); its interpretation as a cost average across items (calculated as if each item was part of a problem with constant conditions, equal to the local conditions for the item) is still valid.

In practical cases, a per-item cost estimate can be obtained easily with the following two-step procedure:

(i)   Solve the problem with constant conditions for a representative sample of items and input data,

(ii) Average the solution across all the sampled items to obtain the result.

Note that the cost estimate can be obtained even without defining the decision variables in step (i). Problem 3.5 illustrates the accuracy of the CA method under capacity constraints.

3.6.1 Practical Considerations

While for simple problems, such as the one solved above, the solution can be easily automated, more complex situations may benefit from decision support tools with substantial human intervention. The following two-step human/machine procedure is recommended: (i) first, recognizing that its recommendations may need fine-tuning adjustments, the CA (or other simplified) method is applied to a basic version of the problem without secondary details; (ii) then, trained humans develop implementable solutions that account for the details, perhaps aided by numerical methods that can benefit from the output of the first step.

In some cases, when time is of the essence humans alone may have to carry out this second step because efficient numerical methods capturing peculiar details may not be readily available, and developing them may be prohibitively time consuming. Furthermore, even without time pressures, if the details are so complex (or so vaguely understood) that they cannot be quantified properly, pursuing automation for the fine-tuning step would seem ill-advised. Fortunately, this is not a serious drawback; as argued earlier, significant departures from ideal situations should not increase cost significantly, leaving humans considerable latitude for accommodating details.

As an illustration of these concepts, problem 3.6 re-examines the terminal location problem of Section 3.4. when only 50 specific locations are feasible. The cost of the two-step procedure (fine-tuned by hand) is compared to the ideal cost without restrictions, and (optionally) to the exact optimal cost obtained with dynamic programming. The reader will find that the fine-tuning step often identifies the exact optimum, and when it does not, the difference between the two-step and the exact optimal costs is measured by a fraction of a percentage point. Furthermore, the two-step and one-step (or ideal) costs are very close; of course, provided that n^*_s is not greater than 50.

3.7 Network Design Issues

In all the scenarios discussed so far, the items followed a predetermined path. Real logistics problems, however, often involve the choice of alternative routes (e.g., alternative ways of shipping) between origins and destinations, in addition to the choice of when and how much to dispatch. In some instances one may even be interested in whether certain routes should be provided at all; or even in the design of an entirely new physical distribution network.

We also found in Section 3.1 that there were economies of scale in flow; i.e., the optimal cost per item decreased with D'. Later in this monograph we will have to consider logistics problems with multiple destinations, where an item's route is not predetermined and cost decreases with flow. We discuss here some key features of these problems, and conclude the chapter with a comparison of detailed and non-detailed approaches for logistic system design.

3.7.1   The Effect of Flow Scale Economies on Route Choice

A simple example with one origin and two destinations (see Figure 3.8) effectively illustrates the properties of optimal system designs with and without flow economies of scale. The origin, O , produces items of type i (i = 1, 2) for destination P_i at a constant rate, given by the parenthetical numbers in the figure: $D'_1 = D'_2 = 4$ items per unit time. The combined production rate at the origin is $D'_1 + D'_2 = 8$ items/unit time. The arrows in the figure depict possible shipment trips; these transportation links are numbered 1, 2, 3. While all the items traveling to P_1 , must travel directly between O and P_1 , the items traveling to P_2 may go either directly or via P_1 .

Let us assume that a fraction (to be decided) x , of the items for P_2 are sent via P_1 and the rest are shipped directly. This establishes a flow $x_1 = 4(1+x)$ on link 1 (OP_1), a flow $x_3 = 4x$ on link 3 (P_1P_2) and a flow $x_2 = 4(1 - x)$ on link 2 (OP_2).

We also assume that the total cost on the network can be expressed as a sum of link costs, and that these depend only on their own flows. This is a reasonable assumption if no attempt is made to coordinate the shipping schedules on the three links, as then the prorated cost to each link should be close to the EOQ expression with demand rate equal to the link flow. Thus, if we let $z_i(x_i)$ denote the cost per item on link i when the flow is x_i, the total system cost per unit time is:

With economies of scale, the functions $x_i z_i(x_i)$ increase at a decreasing rate (are concave) as in

$$Total\ cost = \sum_{i=1}^{3} x_i z_i(x_i).$$ (3.16)

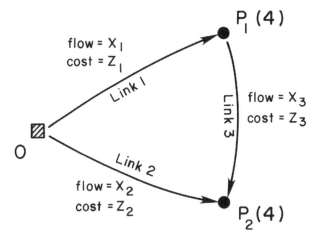

Fig. 3.8   Flows and costs for a simple 3-node network

With economies of scale, the functions $x_i z_i(x_i)$ increase at a decreasing rate (are concave) as in Figure 3.2c. Because the x_i's are linear in the split x , the total cost is a concave function of the split – this (concave) dependence of cost on splits (decision variables) also holds for general networks.

Suppose, for example, that

$$z_1 = x_1^{-1/2}, \quad z_2 = 3x_2^{-1/2} \ and \ z_3 = 1 ;$$
$$x_1 z_1 = x_1^{1/2}, \ x_2 z_2 = 3x_2^{1/2} \ and \ x_3 z_3 = x_3.$$

Then, as a function of x , (3.16) becomes:

$$Total \ cost = 2(1+x)^{1/2} + 6\,(1-x)^{1/2} + 4x. \tag{3.17}$$

This relationship is plotted on Fig. 3.9; as stated, the total cost is a concave function of the split, x . Like any concave function, it reaches a minimum at one of the ends of the feasibility interval. For our data the optimal solution is $x^* = 1$, indicating that everything should be shipped through P_1. The total cost is 6.8. Although shipping everything direct may be better for different data, clearly one would never want to split the flow to P_2 among the two routes $(OP_2$ and $OP_1P_2)$.

A similar "all-or-nothing" principle holds for networks with multiple origins and destinations if the total cost is a concave function of all the link flows (Zangwill, 1968). In that case all the flow from any origin to any destination should be allocated to only one route. This is not difficult to see: one can define a split between any two routes joining an origin and a destination, and since the link flows are linear in that split, the total cost is concave in the split; thus, only one of the routes can carry flow. Networks with diseconomies of scale behave in an opposite manner. In that case the total cost function is convex in the splits and there is an incentive to spread out the flow among routes. In fact, if for a one origin and one destination network, there exist several routes with identical cost functions (with diseconomies); it is not difficult to prove that the total flow should be evenly divided among *all* the routes.

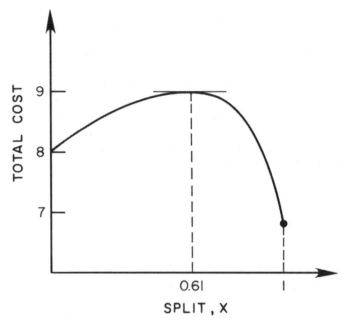

Fig. 3.9   Concave cost function in the split

Networks with flow economies of scale also respond in a different manner to changes in conditions. While, with diseconomies, a small improvement to one of the routes would lead to a small change in the optimal flow distribution (see exercise 3.8), with economies, the optimal flows either stay the same or change by a discrete amount. This can be seen with the example of (3.17). As long as $z_3 < [2 - 2^{-1/2}] \approx 1.3$, x^* equals 1, but if z_3 is increased beyond this value ever so slightly, the solution jumps to $x^* = 0$. This is typical of concave cost problems: minor changes to the input data can induce large changes in the optimal solution. Fortunately, the cost does not behave in such manner; despite the jump in our example the cost is a continuous function of z_3:

$$\text{Total cost}^* = 8 \qquad\qquad if\, z_3 \leq 2 - 2^{-1/2}$$
$$= 8^{1/2} + 4z_3 \qquad if\, z_3 > 2 - 2^{-1/2} \,.$$

Let us now turn our attention to solution methods.

3.7.2 Solution Methods

The nature of the solution is not the only difference between networks with economies and diseconomies; the way to find it is also different. While networks with diseconomies are well behaved optimization problems without local minima that are not global, networks with economies are not. The books by Steenbrink (1974), Newell (1980), and Sheffi (1985) discuss networks with scale diseconomies in detail; Popken (1988) reviews the sparser (traditional/detailed) network design literature for networks with scale economies. Further information on this subject can be found in Ball et al (1995 and 1995a).

Although local search algorithms can be used to find near optimal solutions for large detailed networks with convex costs, the same procedures fail with concave cost networks. The task is then much more complicated, and the network sizes that can be handled by numerical methods much smaller.

Except for technical details, all local search algorithms work in the same manner. First, the total cost is evaluated for an initial feasible solution, described by a set of variables that uniquely identify the decisions; e.g., the set of splits for all origin destination pairs. A small cost-reducing perturbation to the feasible solution (e.g., a differential change to the splits) is then sought. If not found, the search stops because the initial solution is a local minimum; i.e., a solution that cannot be improved without substantial changes. Otherwise, an improved larger perturbation obtained from the original small perturbations is identified, and then used to construct a new improved feasible solution. The process is then iterated (seeking small cost-reducing perturbations to the new solution, etc.) until no significant improvements result.

Local search techniques work acceptably for networks with scale diseconomies, because in those instances any local minimum is a global minimum. Unfortunately, this is not the case with economies of scale. Figure 3.9 reveals that our simple problem has two local minima: $x = 0$ and $x = 1$. If a local search algorithm is applied to our example, any starting solution with $x < 0.61$ (the maximum in the figure) will converge suboptimally to $x = 0$. While for our simple example this can be corrected simply by starting with different x's , the task is daunting for large, highly detailed networks. In that case, the number of potential traps for a local search – all local minima regardless of cost – increases exponentially with the amount of detail.

This is illustrated with an example, where items from a large number N of origins are shipped to one destination using two transportation modes (1 and 2). We use x_i to denote the split of production from origin i sent on mode 1, and assume that (to satisfy an agreement with the providers of type-1 transportation) each x_i must satisfy $x_i \geq h_i$ for some constant $h_i \geq 0$. Transportation by mode 2 is assumed to be more attractive, but limited in capacity; that is, the sum of the x_i's must exceed a value h.

For a set of splits to be feasible, thus, the following must be true:

$$\sum_{i=1}^{N} x_i \geq h, \text{ and } h_i \leq x_i \leq 1, \forall i \qquad (3.18a)$$

We seek the set of feasible splits that minimize the total cost, or equivalently the penalty paid because not all the items can be shipped by mode 2. The penalty paid for each origin is assumed to increase with x_i , except at certain values where a fixed amount ε_i is saved – perhaps because shipments can then be multiples of a box, requiring less handling. To simplify the exposition, let us assume that there is only one such value δ_i for every origin, and that away from this value the penalty equals x_i; otherwise the penalty is $x_i - \varepsilon_i$. If we define $\varepsilon_i(x_i)$ to be: ε_i if $x_i = \delta_i$ and 0 otherwise, then the combined penalty for all the origins can be expressed as:

$$\sum_{i=1}^{N} [x_i - \varepsilon_i(x_i)]. \qquad (3.18b)$$

Note that each one of the terms in this summation for which $\delta_i > h_i$ exhibits two local minima in the range of feasibility $[h_i, 1]$: $x_i = h_i$ and $x_i = \delta_i$.

Any combination of x's, each equaling either h_i or δ_i , and satisfying (3.18a) is a local minimum, which could stop a search. If the δ_i and the h_i are uniformly distributed between 0 and 1, and h is small, there will be $O(2^{N/2})$ local optima. With so many traps, local search algorithms are doomed to failure for this problem – not because (3.18b) is discontinuous, but because it is not convex. A different method must be used.

Certainly, one could search exhaustively over all the possible solutions with a combinatorial tool such as branch and bound, but these methods can only handle problems of small size – typically with $O(10^2)$ decision variables or less.

Alternatively, one could try to exploit the peculiar *mathematical structure* of Eqs. (3.18) – or whichever problem is at hand – to develop a suitable algorithm. If successful, the approach would find a solution with all its detail. In our case, the optimization of (3.18) can be reduced to a knapsack problem that can be solved easily (see exercise 3.8); in other instances it may be possible to decompose the problem into a collection of small easy problems. Very often, however, a simple solution method cannot be found. In our case, this would happen if there were more than one (ε_i, δ_i) for each origin. Traditionally one then turns to ad hoc intuitive solution methods (known as heuristics) which one *hopes* will yield reasonable solutions.

There is also another approach. If while inspecting the formulation, or even better in the process of formulating the problem, one realizes that certain details are of little importance one should leave them out. Our example illustrates how removing minor details can turn a nightmare into an easy problem. If the ε_i's are so small that the $\varepsilon_i(x_i)$ in (3.18b) can be neglected, then the objective function reduces to $\Sigma_i x_i$. Former sources of difficulty, the δ_i and ε_i no longer enter the formulation. With less detail, the problem becomes well behaved (convex), and even admits a closed form solution; e.g., if $\Sigma_i h_i \geq h$ then the optimal splits are $x_i = h_i$ and the total cost is $\Sigma_i h_i = N\overline{h}$.

Note that the optimal cost is given by an average (there is no need to know precisely each individual h_i in order to estimate the optimal cost), and that the optimal solution can be described with the simple rule "make every split as small as possible", which can be stated without making reference to the h_i's.

In the rest of this monograph we will seek solutions to logistic problems using as little detail as possible, describing (as in the example) the solution in terms of guidelines which are developed based on broad averages instead of detailed data. We recognize that the solutions obtained from such guidelines may benefit from fine-tuning once detailed data become available; but also note that incorporating all the details into the model early will increase the effort for gathering data, and, as illustrated, may even get in the way of obtaining a good solution.

Observation of mother nature's logistics networks suggests that many logistics systems can be designed in this manner. Trees can be viewed as a logistic system for carrying nutrients from the soil to an above-ground region (the leaves) to meet the sun's rays. While every individual tree of a species is distinct from other individuals, we also see that the members of a species share many common characteristics *on average*. There is order at the macroscopic level. This is not surprising, since members of the species have adapted to similar environmental conditions, also filling the same niche in the eco-system. The detailed characteristics of an individual tree are (like our logistic systems) developed from two levels of data in two different ways:

(i) Members of the same species share a genetic code, which has evolved in response to the typical or average conditions that can be expected. This code is analogous to the guidelines of a simple model; e.g., "make each split as small as possible."

(ii) In response to the detailed conditions of its location, a tree develops an individuality within the guidelines of the genetic code, better to exploit the local conditions. This would be analogous to the fine tuning that could have taken place if the ε_i, h_i, and δ_i had been given in our example.

The same could be said for other logistic systems encountered in nature, such as the circulating and nervous systems of the human body.

On further inspection we notice that, not only average characteristics, but some *specific* traits are also the same for all individuals, (e.g., some tree species have always one trunk, all humans have one aorta artery, etc.). It is as if nature had decided that these items of commonality are optimal for almost any conditions that can be encountered; therefore, that part of the design is not open to fine tuning. Perhaps the same can be said of logistics systems.

The logistics systems of nature also have economies of scale. It takes less energy to move a certain flow through one single pipe than through two pipes with one-half the cross section. As in our networks with concave costs, there is an incentive to *consolidate* flow into single routes that can handle great volumes efficiently. Nature has responded to this challenge by evolving hierarchical systems of conveyance, such as the three hierarchy network of Fig. 3.10.

Scientists have begun to realize that apparently very complex ("fractal") structures, such as a fern leaf, can be replicated and/or described with just a few rules and parameters (Gleick, 1988, provides an entertaining description of these ideas). For the example of Fig. 3.10 the separations between "nodes" (e.g., A_1 and A_2) for each hierarchy might be found to be relatively constant, perhaps varying with the distance from the root, as might be the number of branches at every node and the relative size of the main and secondary branches at nodes of the same hierarchy. The latter may also vary with the distance from the "root."

A physical distribution network should probably be organized in a similar way with the root becoming the depot, the leaves the customers, and the nodes intermediate transshipment centers or terminals. Physical distribution networks that serve similar purposes, just as in nature, should likely share the same hierarchical organization and overall traits even if the specific details differ. As in nature, it should be possible to describe their near optimal configuration with just a few simple rules and parameters (see problem 3.11).

In this spirit, the chapters that follow will try to get at the "genetic code" of logistics systems; i.e., describe how general classes of logistics networks should be organized, with guidelines for obtaining an optimal structure developed without using detailed data. Building on the simple EOQ model, we gradually consider more complex systems.

Chapter 4 describes problems with a single hierarchy consisting of one origin and many destinations (or the reverse); i.e., "one-to-many" problems. Chapter 5 describes "one-to-many" problems with transshipments (multiple hierarchies), and Chapter 6 concludes with "many-to-many" problems.

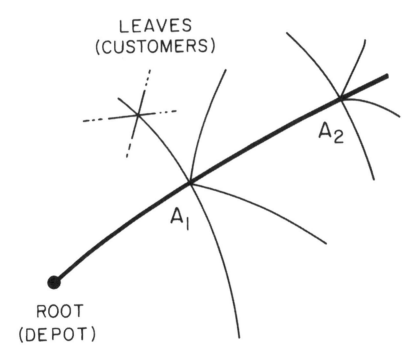

Fig. 3.10   Schematic representation of naturally occurring logistics systems

Suggested Exercises

3.1   Various items have to be shipped from an origin to n destinations, i ,
i = 1, 2, ... , n . Destination i demands v_i truckloads. If v_i is not an integer, a partially full truck will have to be dispatched; trucks are not
allowed to visit more than one destination. The distance traveled by
each truck serving i is d_i . According to Eq. (2.5c) the total cost of
serving i , c_i , can be expressed as:

$$c_i = (v_i)^+ (c_s + c_d d_i), \qquad (3.19a)$$

where $(x)^+$ denotes the smallest integer greater than or equal to x .
Independent of distance, the variable costs $c'_s V$ and $c'_d V d$ are omitted from the expression.

Consider also the approximation:

$$c_i \approx c_{i'} = (v_i + 0.5)(c_s + c_d d_i). \qquad (3.19b)$$

Then:

(i)   Plot c_i and c'_i vs. v_i , and also c_i and c'_i vs. d_i ,
(ii)   For n = 5, c_s = 100, c_d = 1, and the following data:

i	1	2	3	4	5
d_i	100	200	300	400	500
v_i	5.5	8.2	5.7	2.3	1.8

Calculate the exact and approximate total costs, (TC and TC' respectively) and the relative error in the approximation.

Suppose that each predicted value v_i has an error, ε_i , which is the
outcome of a normal random variable with zero mean and standard
deviation, σ = 0.2 ; these variables (for all i) are mutually independent and independent of d_i . Then:

(iii)   If the values of v_i given in part (ii) are the true but unknown
ones, determine the predictions one might expect using Eq.

(3.19a) and Eq. (3.19b). Compare their accuracy relative to the (unknown) truth. You may solve this part either analytically or with a few simulations. Use of a spreadsheet is recommended,

(iv)   Repeat part (iii), analytically for arbitrary c_s, c_d, n and $\sigma < 0.2$, and obtain the expected accuracy of Eq. (3.19a) and Eq. (3.19b),

(v)   Discuss the results, the applicability of Eqs. (3.19a) and (3.19b) in different instances, and the implications for modeling.

3.2   Repeat the CA argument leading to Eqs. (3.11) for the bus dispatching problem outlined at the outset of Section 3.4. Assume that, perhaps due to changes in the prevailing trip purposes with the time of day, c_i is a slow varying function of time $c_i(t)$.

3.3   Formulate the equivalent of Eq. (3.10) for the transit stop location problem discussed in Section 3.4, assuming that every passenger travels from an origin along the line to the final stop. Also assume that pipeline inventory cost cannot be neglected, and that a passenger's riding time is only 1/2 as costly as the same amount of walking time.
    (**Hint:** at the end of the trip the vehicle carries more passengers; therefore, the penalty for stopping is more severe. See Hurdle, 1973, for inspiration). Explain how the formulation can be generalized to passenger trips with arbitrary origins and destinations along the line (see Wirasinghe, 1981, for inspiration.)

3.4   Prove that the best location for a single terminal in Fig. 3.7 satisfies $d/d_{max} = 1-(1-a^0)/[2(1-b^0)]$, where $a^0 = a/d_{max}$ and $b^0 = b/D(d_{max})$. (Hint: Use a graphical argument to demonstrate that, at this position, moving the terminal to either side does not change the access cost).

3.5   Describe a procedure, analogous to the exact method described in Section 3.3.3, for the capacity constrained inventory control problem of Section 3.6. (Hint: As suggested in one of the readings, start with an initial interval between dispatches t_1, and from it construct a sequence of headways that satisfy the capacity constraint and a necessary condition for optimality.) Then for a case with constant demand, develop expressions equivalent to Eqs. (3.15) and evaluate their accuracy as a function of v_{max}.

3.6   Freight is to be exported from a region of variable width, lying on one side of a transportation artery (e.g., a highway or railroad line) that is one thousand distance units (Du) long, L = 1,000 Du's. From the origins freight can only be carried perpendicularly to the artery, unless of course it moves on the artery itself. Freight must flow out of this region through a system of terminals (e.g., ports) that are to be located on this artery. The cost of travel within the region is one monetary unit (Mu) per weight unit (Wu) per Du. The cost of travel beyond the terminals is not considered as part of our study. The freight transportation needs per unit time (Tu) are expressed as a transportation demand density [Wu/(Du*Tu)] which depends on the position along the artery expressed in Du's, x . The demand density, w(x) , is expressed as follows:

$$w(x) \begin{array}{ll} = 5 & if \ \ 0 \le x \le 300 \\ = 1 & if \ \ 300 < x < 700, \ and \\ = 5 & if \ \ 700 \le x \le 1000 \end{array}$$

Then:

(i)   Determine the optimal location and the total access cost per Tu when we locate n = 1, 2 and 3 terminals. (Use the procedure developed in Sec. 3.4),

(ii)   Show which locations you would use if only the attached 50 locations are feasible. Calculate for each case the percent change in access cost,

(iii) If each terminal costs c_T = 80,000 Mu's per Tu to operate (including any relevant amortized fixed costs), determine whether the optimum number of terminals that should be operated, n^*, is 1, 2, or 3. Calculate as well the total regional cost per unit time, including both transportation and terminal operations, z^*. Do the calculation with and without the location constraints described in (ii),

(iv)   Determine n^* and z^* for arbitrary c_T using the continuum approximation (CA) method. Compare the results for c_T = 80,000 with those for part (iii). Discuss how the CA results might differ from the true optimum as $c_T \to 0$, *with and without the 50 location restriction,*

(v)   **Extra credit:** Find the exact optimal solution to part (ii) using dynamic programming. Solve for n = 1, 2, 3, ..., 10. Then determine the ranges of c_T for which the optimum number of terminals is 1,

2, etc. Calculate and plot the resulting z^* as a function of c_T and compare the result with your findings in parts (iii) and (iv).

Data for Problem 3.6

Coordinates of 50 possible locations

19,	37,	46,	54,	77,	79,	90,	129,	132
141,	159,	199,	210,	211,	223,	228,	236,	262
265,	297,	300,	307,	369,	384,	394,	423,	431
473,	475,	529,	543,	549,	551,	572,	640,	646
652,	653,	654,	660,	682,	683,	685,	759,	776
835,	866,	925,	953,	994				

3.7   Suppose that the average link costs for the network of Fig. 3.8 exhibited diseconomies of scale, increasing with flow instead of declining, as follows: $z_1 = x_1$, $z_2 = 3x_2$, and z_3 = constant. Describe then how the optimal split x^* and the optimal total cost change as z_3 is increased from zero to 20,

3.8   Prove that the optimization problem defined by Eqs. (3.18) can be reduced to a knapsack problem if we can assume that a good solution can be found by restricting the x_i to be either $x_i = h_i$ or $x_i = \delta_{i,}$,

3.9   Two factories owned by the same firm export refrigerators (factory A) and stereo components (factory B) through a seaport, C ; the items are taken by truck from A to C and from B to C. A truckload of refrigerators brings in $\$2 \times 10^5$ in revenue and a truckload of stereos ten times as much. The production rates for A and B are D'_A and D'_B truckloads per day, and the inventory carrying cost is 15 percent per annum. The distances d_{AC} and d_{BC} are d_{AC} = 300 miles and d_{BC} = 300 miles; trucks cover 1200 miles/day (drivers are changed), and the transportation cost is $\$1.0$ per truck-mile regardless of the load size being carried (back-hauls have been factored into this cost figure). We assume that production and transportation take place around the clock. Revenues are collected (COD) upon delivery at C,

Ships sail weekly from C and carry all the items that have accumulated at the port when they depart. All the items go to (foreign) port D.

(i) Calculate the optimum dispatching frequency, q , the optimum shipment size, v , $(0 \leq v \leq 1)$ and the resulting cost per day, z, for each manufacturer (use A and B as subscripts for the variables q, v, and z). Plot $z_A(D'_A)$ and $z_B(D'_B)$. Discuss how COD payment upon arrival at D would change the shipping strategy,

Alternatively, COD payments can be received at a land terminal, T, owned by an export company. At this terminal, which is equidistant from A and B $(d_{AT} = d_{BT} = 100)$, items are unloaded and placed onto railcars for cheaper transportation.

The payment received for each truck load is in this case reduced by an amount Δ , which is proportional to the size of the truckload (v) and reflects the cost of transportation from T to C , plus the handling and delay costs at T: $\Delta = 100 \ v \ [1 + D'_T{}^{-1/2}]$. Here, D'_T denotes the number of *full* truckloads handled at T per day.

(ii)   Find the minimum total cost per day as a function of D'_A and D'_B, if all items are sent to T . Assume $D'_A < 0.4$ and $D'_B < 4$,

(iii) If fractions x_A and x_b of the production at A and B are shipped through the terminal, and D'_A and $D'_B = 0.05$ truckloads per day, write the combined minimum total cost per day for both factories z $= z_A + z_B$. It should be a function of x_A and x_B only $(0 \leq x_A , x_B \leq 1)$. You should also assume that shipments from A and B to both C and T take place at regular intervals. Explain your treatment of inventory cost at A and B,

(iv) Consider a local improvement strategy, where the company changes the routing of its shipments if small changes to x_A and x_B result in reduced costs. If they detect an improvement, they keep changing the routing splits (little by little) until no more improvements are possible. Assume that you are in charge and that currently you are shipping everything through the terminal $(x_A = x_B = 1)$. What would happen? Assume that $x_A = x_B = 0$; and then $x_A = 0$, $x_B = 1$. Repeat,

(v)   Draw a picture of the feasible region, sketch some (2 or 3) equicost contours and explain what happens. Can you prove which is the optimum solution?

(vi)   Suppose that A and B are not owned by the same conglomerate. but that each minimizes its own cost. Could you visualize an instance where the minimum total cost solution was not attractive to either A or B?

(vii) Suppose that (due to congestion), $\Delta = 100 \ (x_A + x_B) + 10^4(x_A + x_B)^2$. Repeat parts (iii) to (v).

3.10 The robustness results of Section 3.2 extend to other cost functions, useful in later chapters. Repeat the analysis of Section 3.2 if $z = Av^a + Bv^{-b}$, where $0 < (a, b) \leq 1$. (The appendix in Daganzo and Newell, 1986, discusses the robustness of this expression in the decision variables.)

3.11 Most airline networks have a "hub-and-spoke" structure. Irrigation networks have a "trunk-and-branch" structure.

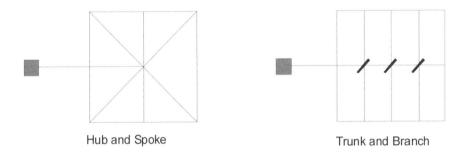

Hub and Spoke   Trunk and Branch

(i) Why shouldn't an irrigation network have a hub-and-spoke pattern? Why don't airlines use trunk-and-branch structures?

(ii) Define qualitatively a network structure (draw picture) that would generalize hub-and-spoke, and trunk-and-branch.

Glossary of Symbols

A:	EOQ formula constant, $A = c_h/D'$,
B:	EOQ formula constant, $B = c_f$,
c_f:	Fixed cost of a shipment,
c_h:	Holding cost per item, $c_h = c_r + c_i$,
c_i:	Waiting cost per unit of time ($/time),
c_r:	Annual rent cost per item ($/item-year),
c_s:	Fixed cost of stopping for a shipment (fixed part of c_f),
c'_s:	Cost of carrying an extra item (fixed part of c_v),
c^0:	Constant,
c'_d:	Transportation cost per unit distance per item,
δ:	Relative error in the EOQ cost coefficients,
Δ:	Variation range of the actual holding cost (unknown) around the EOQ holding cost,
$\bar{d}$:	Depot location on a distance line,
d_i:	Location of terminals on a line segment,
D(t):	Cumulative demand function, as a function of time,
D'(t):	Variable demand rate (as a function of time),
D':	Constant demand rate (items/time),
D'_i:	Production rate of origin O for destination P_i,
$\bar{D}$ ':	Average consumption rate,
D'_{crit}:	Critical value of D' at which the capacity constraint is reached,
d_{max}:	Length of a distance line containing origins,
dt:	Differential of time,
γ:	Relative error in shipment size,
γ':	Ratio of the actual to the optimum cost per item,
h:	Upper bound for capacity of mode 2,
h_i:	Lower bound for x_i,
$H_s(t)$:	Step function defined by $t_i - t_{i-1}$ when t belongs to I_i,
H(t):	Smooth approximation of $H_s(t)$,
I_i:	Generic interval,
$|I_i|$:	Length of the interval I_i,
m_i:	One-dimensional coordinates of M_i,
M_i:	Midpoints of curve segments,
n:	Number of shipments,
N:	Number of origins,
n_s:	Number of stops,
n_s^* :	Optimal number of terminals,

$\Psi\{x\}$: Increasing concave function,

O: Generic origin,

P_i: Generic point; also, generic destination,

Q: Generic point,

R(t):   Receiving step curve for EOQ with variable demand,

s(d):   Continuous function defined by $s(d_i)=m_i-m_{i-1}$,

t_i: Time point,

t_i':   Point belonging to I_i,

T_i: Generic point,

t_{max}: Time horizon,

v: Generic shipment size (items),

v^*:   Optimal shipment size,

$v^{*'}$:   Optimal shipment size with erroneous cost structure,

v_0:   Actual chosen shipment size (instead of the optimum v^*),

v_i: Shipment size of the ith shipment,

v_{max}: Maximum capacity of a vehicle (items),

v_{tot}: Total number of items shipped,

x_i: Flow on link i; also, split of production from origin i sent on mode 1,

x^*:   Optimum split of flows,

z: Objective function of the EOQ model (cost per item),

z^0:   Actual cost per item due to suboptimal v^0,

z^*:   Optimum cost per item in the EOQ model,

$z_h(v)$: Holding cost per item as a function of the shipment size,

$z_i(x_i)$: Cost per item on link i when its flow is x_i,

$z_i^*(n)$: Optimum holding cost obtained by Dynamic Programming.

4 One-to-Many Distribution

Readings for Chapter 4

Because this chapter is based on numerous references, cited in the text, only a representative few are recommended here. The vehicle routing results of Secs. 4.2 and Appendix A are largely based on Daganzo (1984a) and (1984b). A strategy for routing and scheduling deliveries from a single origin to many destinations with similar demand is developed in Burns, et al. (1985); some of the material in Sections 4.3 and 4.6 is related to this reference. Hall (1985) generalizes this strategy for problems with widely different customer sizes; Section 4.7 discusses this problem and similar extensions. Section 4.5, on implementation considerations, is based on two references: Clarens and Hurdle (1975), which shows how a detailed solution to a one-to-many problem can be obtained from CA guidelines, and Robusté, et al. (1990), which discusses ways of fine-tuning such solutions with an automated second step – this reference also compares the two-step approach with detailed optimization methods. Sections 4.6.2 and 4.6.3 discuss planning for uncertainty and dynamic operations; additional ideas along these lines can be found in Daganzo and Erera (1999) and Erera (2000).

4.1 Initial Remarks

This chapter addresses physical distribution problems where items produced at a single origin are to be taken, without transshipment, to a set of scattered destinations over a service region, $\mathbf{R}$. For the most part, the chapter will focus on delivery problems, although it should be recognized that collection problems from many sources to a single destination are mathematically analogous – the theory applies to both. The objective is to obtain simple guidelines for the design of a set of routes and delivery schedules that will minimize the total cost per unit time. The CA approach of Chapter 3 will be extended to this problem; yielding in the process simple formulae for the total resulting cost.

We saw in Chapter 3 that the continuum approximation method is most accurate for one-dimensional point location problems if the characteristics of the problem vary slowly along the location domain (e.g., the time or distance line). Although our current problem is much more complex – in addition to a schedule for every customer, we must design a set of time-varying routes to meet the schedule – it can be reduced to a point location problem in multiple (time-space) dimensions; accordingly, our solutions will be most accurate if the characteristics of the problem vary slowly over both space and time. An outline for this chapter is given after explaining this assumption in more detail.

The results in this chapter apply to situations where a large number of destinations/customers N is distributed over a region $\mathbf{R}$ in a form that can be described by a slow varying continuous density function $f(\mathbf{x})$ of the point coordinates $\mathbf{x} = (x_1, x_2)$ within $\mathbf{R}$. That is, the actual number of points in a subregion of $\mathbf{R}$, $\mathbf{A}$, is approximately given by:

$$\int_{x \in A} Nf(x)\, dx.$$

If $f(\mathbf{x})$ remains nearly constant over $\mathbf{A}$, ($\mathbf{A}$ is small), then the number can be written as:

$$N \int_{x \in A} f(x)\, dx \cong Nf(x_0)A,$$

where $\mathbf{x}_0$ is any point in $\mathbf{A}$, and A is the area of $\mathbf{A}$. Note that a design approach based on expressions of this type can be used even *before* the actual point locations are known. In the literature, a common interpretation of $f(\mathbf{x})$ is as a probability density function for the coordinates of the customers, assumed to be located independently of one another. In that case, the above expressions represent the mean number of customers found in subareas of $\mathbf{R}$; the actual number can vary across subareas with the same mean. Because one also finds that the standard deviation grows with N and A more slowly than the mean – it is of the form $\{Nf(x_0)A [1-f(x_0)A]\}^{1/2}$ when $f(\mathbf{x})$ is nearly constant over $\mathbf{A}$ and points are located independently – these variations do not prevent continuous approximations to improve as N grows.

Finally we also assume that the cumulative number of items demanded by each customer can be expressed as a demand curve, $D_n(t)$ (n = 1, 2, ..., N), which is assumed to vary slowly with t.

The chapter addresses one-to-many design problems, starting with the simplest situations and gradually incorporating complicating features. The analysis uses some basic results for the vehicle routing problem, which are described in Sec. 4.2. Sections 4.3 to 4.5 examine in detail a special case where all the customers are identical, $D_n(t) = D(t)$, and show how (for that case) the continuous approximation method can be used to design a rational service strategy and to obtain cost expressions. Section 4.3 develops guidelines and cost expressions for situations where the vehicles are filled to capacity at the depot, section 4.4 analyzes (undesirable) situations when they cannot and section 4.5 shows how a detailed solution can be constructed from the guidelines.

Sections 4.6 and 4.7 examine the case with different $D_n(t)$ across n . Section 4.6 shows how to extend the guidelines and cost formulas for identical customers to this case, assuming that all the customers are treated alike; strategies like these are called "non-discriminatory" or "symmetric". Asymmetric strategies, useful when the customers are very different, are examined in Section 4.7.

Up to Sec. 4.7, it is assumed that the production process at the origin can be adjusted without a penalty to meet the scheduled shipment quantities. Section 4.8 relaxes this assumption and discusses a general method to deal with other peculiarities.

4.2   The Transportation Operation

This section describes the transportation operation and simple formulas that capture its performance. It first shows that, given a set of delivery schedules for the customers in the region, one should use the vehicle routes which minimize total distance traveled – the main determinant of transportation cost. The bulk of the section (Secs. 4.2.1 and 4.2.2) then presents distance formulas for certain non-detailed, near-optimal vehicle routing strategies. The results are used throughout the chapter to determine schedules, and then again throughout the remainder of the monograph.

It is assumed throughout the chapter that items are distributed with identical vehicles capable of carrying v_{max} items. This definition of vehicle capacity can be used even if different item types move through the system, simply by redefining the concept of "item". If the maximum freight volume (or weight) that can be carried by a vehicle does not depend appreciably on the mixture of item types making up its load, one can think of an "item" as a unit of volume (or weight) and of v_{max} as the vehicle's volume (or weight) capacity. Each destination can then be viewed as a consumption center for packages of unit volume (or weight) – "items" – containing an appropriate product mixture. Section 4.8 discusses more complex capacity restraints and suggests further readings.

Vehicles are dispatched on service routes from the origin (depot) at times t_1, t_2, etc., on delivery runs to particular subsets of customers (possibly the entire set each time). Since vehicles are identical, an operating strategy can be defined relatively easily. We seek the set $\{t_\ell\}$, as well as the delivery lot sizes and the specific customers served each "ℓ" ; i.e., the delivery schedules for every customer.[1] We also seek the routes that minimize transportation cost at each t_ℓ . Our task is simplified because, as shown below, the combined length of all the routes is the main determinant of cost, and simple route length formulas exist.

We saw in Chapter 2 that the cost of transportation on one vehicle route from one origin to several destinations was approximately a linear function of the total size of the shipment, the number of stops and the total distance traveled. If costs on all the vehicle routes are additive (this seems reasonable), the cost of serving all the destinations for time t_ℓ should be the sum for the costs on each route; i.e., a linear increasing function of the total number of routes (vehicles) used, the total volume shipped, the total number of stops, and the total distance. For a given set of delivery schedules to each destination the total volume shipped at each t_ℓ is obviously fixed. Thus, we only need to focus on the number of routes, delivery stops, and vehicle-miles when seeking delivery routes for time t_ℓ.

We consider only strategies that minimize the number of delivery stops by avoiding to the extent possible customer load-splitting among vehicles. This is achieved if each destination is visited by the minimum possible number of vehicles able to hold its delivery; i.e., one vehicle if $v < v_{max}$ items are to be delivered, and $[v/v_{max}]^+$ otherwise.[2] (For customers receiving $v > v_{max}$ items, one would dispatch $[v/v_{max}]^-$ full vehicles exclusively to the customer, and would consolidate the remaining items with smaller deliveries to other nearby customers on a single vehicle route.) Although in some instances it may be possible to reduce the number of tours and the distance traveled by splitting loads (see problem 4.1), the reductions are unlikely to be significant in most cases. Of all the possible strategies without load-splitting we prefer the one with the least distance, as this strategy should also minimize the number of vehicle routes. (This should be intuitive; a set of routes which minimize total distance should use vehicles to the fullest because fewer line-haul trips to and from the depot then need to be made.)

Since a reasonable set of vehicle routes can be chosen on distance grounds alone, the routes can be designed without knowing the magnitude

---

[1] In this chapter "ℓ" will be used to index dispatching times and headways, and "i" to index service districts.

[2] We denote by $[x]^+$ the smallest integer greater than or equal to x; and by $[x]^-$ the integer of part of x.

of the cost coefficients in Eq. (2.5d). Focusing on the difficult case when v is smaller than v_{max}, the remainder of this section discusses distance-minimizing routing schemes and presents simple formulas for estimating distance (and therefore transportation costs). The results depend on the number of customers to be served at time t_ℓ, their spatial distribution in the region, and on the number of stops that vehicles can make C = $[v_{max}/v]$. It is assumed that the lots carried to each customer are of similar size (a reasonable assumption for the cases with identical customers discussed in the first few sections of this chapter), so that C is the same for all vehicles; in later sections, C will be allowed to vary.

4.2.1   Nondetailed Vehicle Routing Models: Many Vehicle Tours

Eilon, et al. (1971), developed simple approximate formulas for the distance of near-optimal vehicle routes in Euclidean square regions. The discussion presented here is based on more recent material extending Eilon et al.'s results to zones of arbitrary shape (Daganzo, 1984a and 1984b), and also incorporating other metrics and the influence of underlying transportation networks (Newell and Daganzo, 1986 and 1986a, and Newell, 1986). Appendix A summarizes the logic behind some of these results.

In order not to introduce additional notation, we will use N to denote the number of destinations that must be visited. If tours are not being constructed for all the customers in the region, as occurs later in the chapter, the results can be easily reinterpreted.

We have already mentioned that vehicles should be used to the fullest; that is, there should be at most one vehicle that makes fewer than C stops, and none if N is an integer multiple of C . Our strategies are of the "cluster-first and route-second" type, where the service region is divided into non-overlapping zones of C customers, to be served by separate vehicles. For a given set of zones, the vehicle routes are easy to construct using some simple rules. To minimize the total distance (and, hence, the cost), these zones should have specific shapes and orientations, dictated by the relative magnitude of N and C^2 . Two cases need to be considered: (i) when the number of vehicle routes N/C is much greater then the number of stops per route C , $N >> C^2$, and (ii) when only a few vehicle routes are needed $N << C^2$.

For case (i), discussed in this subsection, delivery districts (or zones) should have a width comparable with the distance between neighboring points and be as long as necessary to contain C points; see Appendix A. The formulas are most transparent when expressed in terms of the spatial point density – in points per unit area – evaluated at a point inside the delivery district, $\mathbf{x}$: $\delta(\mathbf{x}) = Nf(\mathbf{x})$. (Because $\delta(\mathbf{x})$ varies slowly, just like $f(\mathbf{x})$, it does not matter which $\mathbf{x}$ is used). The factor $\delta(\mathbf{x})^{-1/2}$, appearing in the formulas, represents a distance close to the average separation between neighboring points in the vicinity of $\mathbf{x}$. For randomly scattered points, it has been found that (see Appendix A):

$$(zone\,\text{width}) \approx (6/\delta)^{1/2}$$
$$(\text{zone length}) \approx C/(6/\delta)^{1/2}.$$

These dimensions are close to ideal and relatively independent of the metric or underlying network. When δ changes over $\mathbf{R}$, district dimensions should also change over $\mathbf{R}$, although more slowly. As the solution to the EOQ problem, these expressions are robust; departures from the ideal dimensions by 20 - 30 percent are largely inconsequential, but larger departures increase distance. This robustness makes it easy to carve out $\mathbf{R}$ into delivery districts of satisfactory dimensions.

Zones should also be oriented "toward the depot", but the precise meaning of this recipe depends on the underlying metric. One should build equidistance contours from the depot and design zones of the right dimensions that are perpendicular to these contours. For the Euclidian metric the contours are concentric circles centered at the depot, so that the zones should fan out from the depot in the radial direction. For the L_1 (or "Manhattan") metric,[3] the contours are squares centered at the depot, at $45°$ to the metric's preferred directions; in this case the zones should be perpendicular to these contours, so that they don't point *exactly* toward the depot. Ideal orientations can also be defined when the network includes fast/cheap roads.

Because the zones are narrow, it is easy to construct good vehicle routes, once the region has been carved into delivery districts. One simply needs to travel up one side of the zone, visiting the points in order of increasing distance to the depot, and then return along the other side visiting the remaining points in the reverse order. The effectiveness of this routing scheme improves with the slenderness of the zones – it is exact if zones are infinitely narrow. This has been verified experimentally albeit indirectly by Robusté et al. (1990).

Before turning our attention to distance formulas, let us show how to partition a region into delivery distances with proper shape and orientation. We recommend drawing delivery zones around the region's edge away from the depot, and then filling in the remaining space with more delivery

---

[3] The L_1 distance between two points is the sum of the absolute differences in their coordinates.

routes, always proceeding toward the depot. Figure 4.1 depicts an intermediate point of this process for an irregular region with an internal depot and a rectangular grid network – note how most districts are perpendicular to the (square) equi-distance (L_1) contours. As we progress toward the depot, it may become necessary to pack a few zones with the "wrong" shape, but most will have the right dimensions and orientation. Because the distance traveled is not overly sensitive to (small) deviations from the ideal design, the distance formulas about to be developed should be accurate. This is confirmed by experiments in Daganzo (1984b), Robusté et al. (1990) and Hall (1993).

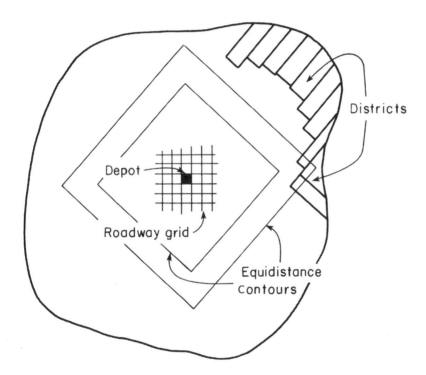

Fig. 4.1   Intermediate stage of the delivery district design process

The last of these references considered systems in which the number of stops in a tour depends on the shipment sizes handled at each stop.

The total distance traveled to visit the C points in a given zone containing point x_0 is:

$$Tour\ distance \approx 2r + \left[k\delta^{-1/2}(x_0)\right]C, \tag{4.1}$$

where $\bar{r}$ is the average distance from the C points to the depot (on the shortest path) and k is a dimensionless constant that depends on the metric ($k \approx 0.57$ for the Euclidean metric, and $k \approx 0.82$ for the L_1 metric). See Appendix A for more details.

The first term of (4.1) can be interpreted as the *line-haul* distance needed to reach the center of gravity of the points in the zone, and the second term as a *local distance* that must be traveled because the points are not next to one another. Note that each stop contributes toward the total a distance comparable with the separation between neighboring points, $k\delta^{-1/2}(x_0)$. This occurs, because the vehicle must be detoured on every leg between successive deliveries. In actuality, because there are only C - 1 such legs, the factor "C" in (4.1) should be replaced by "C - 1" . Thus, a better expression is:

$$Tour\ distance \approx 2\bar{r} + [k\ \delta^{-1/2}(x_0)][C - 1]. \tag{4.1a}$$

The improvement afforded by this expression, particularly obvious for $C = 1$, fades in importance as C grows. Because (4.1) is more compact, it will be used unless C is small.

Let us now see how the total distance over **R** can be expressed without regard to the detailed position of points, using a continuum approximation. Distance (4.1) can be prorated to each one of the points in the zone so that if point i (located at x_i) is r_i distance units away from the depot, then:

$$\begin{aligned} Distance\ for\ x_i &\approx \frac{2r_i}{C} + k\ \delta^{-1/2}(x_0) \\ &\approx \frac{2r_i}{C} + k\ \delta^{-1/2}(x_i) \end{aligned} \tag{4.2}$$

where the second approximate equality follows from the slow varying property of $\delta(x)$.

The total distance traveled in the region is the sum of (4.2) across all points:

$$Total\ distance \approx \frac{2}{C} \sum_i r_i + k \sum_i \delta^{-1/2}(x_i). \tag{4.3}$$

For large N , the summations can be replaced by integrals *independent of the specific location of all the points:*

$$\sum_i \delta^{-1/2}(x_i) \approx \int_R \left[\delta^{-1/2}(x)\right]\delta(x)\,dx, \qquad (4.4a)$$

and

$$\sum_i r_i \approx \int_R r(x)\delta(x)\,dx. \qquad (4.4b)$$

Thus, (4.3) can be rewritten as:

$$Total\ distance \approx \int_R \left[\,2r(x)/C + k\,\delta^{-1/2}(x)\right]\delta(x)\,dx. \qquad (4.4c)$$

Note that this expression is well suited for continuum approximations because the cost in any given (small) area only depends on the local conditions.

An alternative expression for the total distance is obtained after replacing $\delta(x)dx$ by $Nf(x)dx$ in (4.4a and 4.4b); it then becomes clear that these expressions can be interpreted as the product of N and the expectation of $r(x)$ or $\delta^{-1/2}(x)$, when the probability density of position is $f(x)$. Thus, letting $E(r)$ and $E(\delta^{-1/2})$ denote these expectations, the total distance can be expressed as:

$$Total\ distance \approx \left\{\frac{2E(r)}{C} + k\,E\!\left(\delta^{-1/2}\right)\right\} N. \qquad (4.5a)$$

For a uniform density, $E(\delta^{-1/2}) = \delta^{-1/2} = \sqrt{|R| / N}$ and we can write:

$$Total\ distance \approx \frac{2E(r)}{C}N + k\sqrt{|R|\,N}, \qquad (4.5b)$$

where $|R|$ denotes the surface area of R .

Independent of the specific locations, Eqs. (4.4) and (4.5) are particularly useful if cost must be estimated before the point locations are known. In that instance, it may be reasonable to view the actual locations (x_1 , ... ,

x_N) as outcomes of i.i.d random variables with density $f(x)$, and interpret Eq. (4.5a) as the average total distance over all possible locations (x_1 , ... , x_N) . In any specific instance there will be some discrepancy between (4.5a) and the actual distance – for large N most of the difference typically will arise from fluctuations in $\Sigma_i r_i$, which are of order $O(N^{1/2})$ and comparable to the second term of (4.5a). If more accuracy is desired, one should wait for the point locations to become known. Comparisons made in Hall et. al. (1994) indicate that the approximation formulas just presented are fairly accurate even if the number of stops is not the same for all tours. That reference examines improved routing methods for problems in which the number of vehicle stops depends primarily on the vehicle's capacity and the shipment sizes handled at each stop.

4.2.2   Non-Detailed Vehicle Routing Models: Few Vehicle Tours

If $C^2 \gg N$, the optimal strategy must be different from the one we just explained because zones of ideal length (approximately $C/(6\delta)^{1/2}$) would be too long to fit in the service region.

It is not too difficult, however, to design a partition of the region that will yield a distance close to a lower bound for the optimum; i.e., a near-optimal partition. The lower bound is the distance for the shortest single tour visiting all the points, beginning and ending at the depot – the "traveling salesman problem" (TSP) tour. Before describing the partitioning strategy, however, we must introduce some basic properties of TSP tours with many points.

It is well known in the TSP literature (Karp, 1977, and Eilon et al. 1971) that if a region with a nearly constant density of points is partitioned into a few subregions with many points each, then *the length of the shortest tour in the region is close to the sum of the optimal subregional tours.* Appendix A contains a simple proof. The result should be intuitive because: (i) a grand tour can be constructed by connecting the optimal tours of the subregions with a few new legs, while at the same time deleting a like number of existing legs; and (ii) subregional tours can be constructed from a grand optimal single tour, by connecting the broken sections of the grand tour within each subregion with legs along its boundary. In both cases – (i) and (ii) – the original (optimal) and modified (suboptimal) tours differ in total length by no more than the combined perimeter of all the subregions, which is a relatively small quantity when the number of points is large. Thus, the optimal grand tour should be just about as long as all the optimal subregional tours combined.

This property suggests that if the density of points is constant, then the

TSP tour for a subregion $1/4^{th}$ the region's size (with $1/4^{th}$ the points) should be about four times shorter; that is, the average distance per point should be roughly constant. Since the only distance parameter of the problem is $\delta^{-1/2}$, the distance per point for large N must be of the form: $k'\delta^{-1/2}$, where k' is a dimensionless constant, independent of region shape but dependent on the metric; k' is believed to be about 0.75 for the Euclidean metric with randomly distributed points. The expression also holds, with a different k', for regular arrangements of points. Note that the total tour distance can be expressed as: $k'[N|\mathbf{R}|]^{1/2}$. In light of the TSP tour partitioning property, it should not matter much how the region is partitioned for the vehicle routing problem (VRP), provided that travel external to the districts is avoided by ensuring that every zone touches the depot. In that case each VRP tour will be similar to the TSP in the district (the TSP may not have to visit the depot), and the combined VRP length should be close to the overall TSP length; i.e., the lower bound. This means that traditional "sweep"-type algorithms for the VRP, which result in wedge shaped districts as we desire, should work well for the case with $N \ll C^2$.

Alternatively one can build a TSP for the whole region, $\mathbf{R}$, and partition it into segments of C points each that would be connected to the depot. The length of these segments is negligible compared to the total (if $N \ll C^2$), so that the length of all the tours should be close to the length of a TSP. In either case, the length of all tours is close to the TSP lower bound. If the density is constant, we can write:

$$Total\ distance \approx k'\,N\,\delta^{-1/2} = k'\sqrt{N\,|\,\mathbf{R}\,|}\,. \tag{4.6a}$$

As an aside we note that Eqs.(4.5) and (4.6a), which rest on partitioning properties of TSP and VRP tours, may need to be modified for systems in which the distance metric cannot be used to define a "norm". An example of such a metric is a rectangular warehouse with a system of transversal aisles that block travel in the longitudinal direction, except along the sides of the rectangle. For this type of system the length of a tour in which all the aisles with one or more service points are traversed in succession is (Kunder and Gudeus, 1975):

$$Total\ distance \approx 2y_1 + ay_2$$

where y_1 and y_2 are the longitudinal and transversal dimensions of the rectangle, and "a" is the number of aisles containing a point. (Hall, 1993a, has refined this routing scheme and given improved formulae for "warehouse routing".) If N is so large that each aisle contains many points it should be clear that: (i) the traversal strategy becomes optimal and (ii) the coefficient "a" of the above expression can be replaced by the number of aisles.

This shows that the above expression is not of the form (4.6a) since both its terms are independent of N for N →∞.[4]

We now return to the (usual) cases where (4.6a) can be applied, and note that for slow-varying nonuniform densities this distance expression can be approximated by the sum of expected TSP lengths over subregions with many points and (nearly) constant density. In integral form this is:

$$Total \ distance \approx k' \ N \ E\left(\delta^{-1/2}\right), \tag{4.6b}$$

where $E(\delta^{-1/2})$ is given by (4.4a). The uniform density can be shown to maximize (4.6b); thus Eq. (4.6a) is an upper bound to (4.6b).

Notice that, unlike Eqs. (4.4c and 4.4a), Eqs. (4.6) are independent of C; i.e., if vehicles make so many stops that zones of ideal length cannot be packed in the service region, then travel distance is not decreased appreciably by increasing C.

The vehicle routes within the wedge shaped zones are more difficult to develop in this case than in the previous one, which should not be surprising since the TSP problem is NP-*hard*. Nonetheless, simple algorithms such as the ones described in Daganzo (1984a) and Platzman and Bartholdi (1989) can yield tours within 20 percent of optimality. Simple fine-tuning corrections (see Newell and Daganzo, 1986) can then reduce its length by another 10 or 15 percent. Other fine-tuning approaches can yield tours even closer to optimality (see Robusté et al. 1990). It is not our purpose to describe here existing tour construction methods, since this is of marginal value for the theories that will follow. Suffice it to say that, in practice, it is possible to obtain tours within a few percent of optimality with an effort that only grows *proportionately* with the number of points to be visited.

This concludes our review of VRP models and we can now return to the one-to-many problem with identical customers. Recall that we are seeking the set of delivery schedules for each customer and that, given the schedules, the transportation cost at each t_ℓ can be easily estimated with the results that have just been presented. The chosen schedule should strike the best balance between transportation and holding costs.

---

[4] Something similar happens for the VRP. If a set of VRP tours is built by partitioning the single tour through the warehouse, then the total length of the line-haul connections should be 2E(r)N/C (Daganzo and Newell, 1987). This means that the first term of (4.5b) still applies to "warehouse tours", but its second (local travel) term must be replaced by $(2y_1 + ay_2)$.

4.3   Identical Customers; Fixed Vehicle Loads

This section considers strategies where the loads carried by each vehicle are given. Since one would then operate the smallest possible vehicles able to carry the loads, we will denote by v_{max} the load size used. The next section will relax this assumption.

Given $D_n(t) = D(t)$ for t in $[0, t_{max}]$, we seek the dispatching times $\{t_\ell: \ell = 0,...,L\}$ and vehicle routes which minimize the total logistic cost. We let $t_0 = 0$ and $t_\ell \leq t_{\ell+1}$. Because all the customers are alike, there is no compelling reason to treat some differently from others, and we shall assume that every customer is visited with every dispatch, ℓ . Under these conditions, the search for the t_ℓ is facilitated considerably because, as is shown below, the transportation cost only depends on the number of dispatches, L.

Decomposition principle: We now show that for a given number of dispatches, L , the total transportation cost between t = 0 and t = t_{max} is independent of the headways: $H_\ell = t_\ell - t_{\ell-1}$ ($\ell = 1$, ... , L).

We have already stated that the transportation cost for a given ℓ is a linear function of the number of routes, the total number of delivery stops, the total number of items carried and the total distance (see Eq. (2.5d)). Clearly, the combined cost for all ℓ must also be a function of these four descriptors. Because vehicles travel full, three of these (the total number of items $D(t_{max})N$, the number of vehicle tours $D(t_{max})N/v_{max}$, and the total number of delivery stops NL) are fixed; they do not depend on when or how much is shipped at each t_ℓ .

For a given L , the total combined distance for all dispatches is also independent of the t_ℓ . As indicated by Eqs. (4.5), it is the sum of a local distance term proportional to the total number of stops made NL, $kLNE(\delta^{-1/2})$, and a line-haul component which is proportional to the (fixed) number of vehicle tours: $2E(r)(\text{\# tours}) = 2E(r)D(t_{max})N/v_{max}$. Note that the line-haul component is independent of L .

With the cost coefficients defined for Eq. (2.5d), the total transportation cost between t = 0 and t = t_{max} is:

$$\begin{pmatrix} Total\ combined \\ transport\ cost \end{pmatrix} \approx c_s\ N \left\{ \frac{D(t_{max})}{v_{max}} + L \right\} + c_d\ kLNE(\delta^{-1/2})$$
$$+\ c_d 2E(r)D(t_{max})N/v_{max} + c'_s\ D(t_{max})N, \tag{4.7}$$

which only depends on one decision variable, L . An expression based on Eqs. (4.6) instead of (4.5) would be quite similar, and also independent of the $\{t_\ell\}$. Note that Eq. (4.7) *holds regardless of how many items are in-*

cluded in each shipping period, ℓ – even if customer lot sizes are greater than v_{max} (see Problem 4.1). It holds in particular if one decides to ship larger quantities than necessary in anticipation of future increases in the demand curves. This has a profound implication for inventory control. Given a number of shipments L to be received by a customer, their sizes and timing can be chosen to minimize holding cost without affecting the transportation cost. This is explained next.

4.3.1 Very Cheap Items: $c_i \ll c_r$

We examine first a case where items are so cheap (c_i is small) that most of the holding cost arises because of the rent paid to hold the items, $c_h \approx c_r$. In future sections, with more expensive items and different customer types, the CA approach will be used to solve this problem. This is not possible now because, since the rent cost is a function of the *maximum* inventory held, said cost cannot be prorated to (small) time intervals based only on the inventories held at those times. Fortunately, for a given L the transportation cost is fixed, and the headways only influence the rent cost. Clearly, the headway selection problem is analogous to that examined in Sec. 3.3.1.

We saw in Section 3.3.1 that holding cost is minimized if all shipments are just large enough to run out before the next delivery; and that if rent costs were the dominant holding costs (so that the rent cost was proportional to the maximum lot size) then one should choose the dispatching times so as to minimize the maximum lot size. As shown in Fig. 3.3, all the lot sizes should be equal, and given by $D(t_{max})/L$. The same occurs now. The minimum holding cost (for L dispatching periods) is thus:

$$\text{Combined holding cost} = N \left(\frac{D(t_{max})}{L} \right) c_r t_{max}. \qquad (4.8)$$

The sum of this expression and (4.7) yields the total combined logistic cost. The optimal number of dispatching times, L, should be chosen by minimizing such a sum. Only the first and second terms of (4.7), capturing the local stop cost and the local distance cost, depend on L. The other terms, corresponding to the line-haul travel and the loading/handling cost, do not. Thus, the optimal L is the solution of an integer constrained EOQ equation that balances the local transportation cost and the rent cost; the solution is close to:

$$L \approx \left[\frac{c_r \, t_{max} \, D(t_{max})}{[c_d + c_d \, k \, E(\delta^{-1/2})]} \right]^{1/2},$$

(4.9)

if this quantity is greater than 1. Then we can write:

$$\left(\begin{array}{c} \text{Total combined} \\ \text{cost per item} \end{array} \right) \approx [c_s + 2c_d \, E(r)]/ v_{max} + c'_s +$$

$$+ 2\{c_r [c_s + c_d \, k \, E(\delta^{-1/2})] / \overline{D'}\}^{1/2},$$

(4.10)

where, as before, we use $\overline{D'}$ for the average demand rate per customer, $D(t_{max})/t_{max}$. Remarkably, the optimal cost does not depend on the shape of $D(t)$. Not many details are needed to provide a reasonable estimate of operating cost.

4.3.2   *More Expensive Items: $c_i \gg c_r$*

As we did in Chapter 3, we now discuss the problems for items so expensive per unit volume that most of the holding cost is inventory cost. That chapter showed how a CA approach could be used to locate points on the time line (the delivery times) in order to minimize approximately the sum of the holding and motion costs.

The latter was modeled by a constant, c_f , that represented the added cost of each dispatch. Reasonable for the one-to-one problem examined at the time, this simple formulation also applies now; note from Eq. (4.7) that with each additional dispatch, the transportation cost still increases by a constant amount,

$$c_f \approx \left[c_s + c_d \, k \, E(\delta^{-1/2}) \right] N.$$

(4.11)

This constant represents the local transportation cost induced by the N additional customer visits resulting from the extra dispatch. The line-haul cost remains unchanged.

Consequently, the results and methods of Chapter 3 for the EOQ with variable demand also apply here if one defines c_f with Eq. (4.11), and replaces $D(t)$ by $ND(t)$. Equations (3.11) can then be used to estimate cost. Of course, one should remember to add the (large) fixed components of Eq. (4.7) that do not depend on L.

Once the dispatch times $\{t_\ell\}$ and the corresponding delivery lot sizes $\{v_\ell\}$ have been determined, the vehicle routes can be designed as described in Section 4.2, recognizing that the number of stops per vehicle ($C = n_s^\ell \approx v_{max}/v_\ell$) changes with ℓ.

For the special case with uniform density and constant demand, the cost formula reduces to a form analogous to Eq. (4.10), with c_i, D' and $(|\mathbf{R}|/N)^{1/2}$ substituted for c_r, $\overline{D'}$ and $E(\delta^{-1/2})$ (Burns et al. 1985). This approach has been used to streamline General Motors' finished product distribution procedures. The results have been compared with those of (less efficient) direct shipping strategies (Gallego and Simchi-Levy, 1988.)

4.3.3 Inventory at the Origin

The theory we have described focused on the holding cost at the destination and used cost expressions as if there were an equivalent cost at the origin. This assumption, reasonable for the one-to-one problem, is now shown also to be reasonable if the one-to-many system is operated as we described. As we shall see, however, a modification to the operating procedure can drastically reduce the origin holding costs.

With our dispatching strategy, where all the destinations are served with each ℓ, the number of items accumulated at the origin reaches a maximum immediately before a dispatch, and at the destinations immediately after a reception. If production is flexible, one will produce by dispatch ℓ only those items that must be sent by time t_ℓ (and no more) ; thus, the maximum accumulation at the origin is the size of the largest shipment received by any customer, times N. Because shipments arrive as supplies run out, this is also the maximum accumulation for all the customers. It is thus reasonable to represent rent cost by the product of a constant, c_r, and the maximum accumulation, as we have done.

Inventory costs are slightly different. If one could produce the items as fast as desired, one would produce each combined shipment, ℓ, during a short time interval prior to t_ℓ ; and would therefore avoid inventory costs at the origin. This is not likely, however. Although the production rate can change with time to satisfy a slow varying demand $D(t)$, items are normally produced at a roughly uniform rate during each inter-dispatch interval, since most production processes benefit from a smooth production curve. Thus, inventory costs should not be reduced in this manner. If some destinations request more expensive items than others, then inventory cost may be reduced without altering the production rate, simply by changing the order of production; one might want to produce the cheap items at the beginning of the inter-dispatch interval and the most expensive at the end. In most cases, however, only a fraction of the inventory cost at the origin could be saved by exploiting these differences.

Thus, the waiting cost at the origin should be comparable to the waiting cost at the destinations, and a strategy which assumes that both holding costs are equal should yield costs close to one which recognizes the inventory cost at the origin more accurately. (Remember from Chapter 2 that an error in a cost parameter by a factor of 2 only increases the resulting EOQ cost by about 10 percent).

Staggering production for delivery regions: With our operating strategy, all the points in the region, **R** , are visited at each instant, ℓ . However, if instead of waiting for time t_ℓ , vehicles are dispatched just as soon as their last item is produced, both the storage room and the inventory cost at the origin may be reduced. As shown below, this reduction is largest if one can produce all the items for each one of the delivery districts, in sequence.

If the delivery times to any customer are shifted by a time Δt_ℓ smaller than one headway (i.e., the new delivery times are $t'_\ell = t_\ell - \Delta t_\ell > t_{\ell-1}$) , and if Δt_ℓ changes slowly with ℓ so that the new headways are close to the old, then the total holding cost does not change appreciably. With a slow varying D(t), the maximum accumulation remains virtually unchanged, and so does the total number of items-hours; see the difference between the solid and dotted R(t) curves in Fig. 4.2. This is consistent with the CA solution; the cost is sensitive to the delivery headways used as a function of time but much less so to the specific dispatching times.

Suppose that we label the tours used for the ℓth shipment: j = 1, 2, 3, etc. Assume that items for destinations in tour j = 1 are produced first, items for destinations in j = 2 second, etc; and assume as well that every tour is started as soon as the orders for its customers have been completed. If the delivery districts do not change with every ℓ it would be possible to label them consistently so that all destinations would have the same label in successive dispatches. This would ensure that the ℓth delivery headway to every customer is close to $(t_{\ell+1} - t_\ell)$, and that as a result the holding cost at all the destinations would remain essentially unchanged. The ordered production schedule, though, would cut the maximum and average inventory at the origin by a factor equal to the number of tours used for the ℓth shipment, drastically reducing holding costs at the origin.

Unless the demand is constant, $D(t) = \lambda t +$ constant, it is not reasonable to assume that all the delivery districts remain the same; in that case a less ambitious version of our staggered production schedule can be employed.

The service region can be partitioned into production subregions P_1, P_2, ... P_P , where P is a number small compared with the number of tours in any ℓ , but significantly larger than 1 (so that it can make a difference.)

Each production subregion should contain the same number of customers (i.e., the same total demand) and require at least several tours to be covered. Under such conditions, the distance for covering **R** with a VRP is not much different from the collective distance of separate VRP's to cover **P**₁ , wait.

Each production subregion should contain the same number of customers (i.e., the same total demand) and require at least several tours to be covered. Under such conditions, the distance for covering **R** with a VRP is not much different from the collective distance of separate VRP's to cover $\mathbf{P}_1$, $\mathbf{P}_2$, etc. This is true because, like the TSP, the VRP exhibits a partitioning property. (This should be obvious from the material in Sec. 4.2 since: (i) the cost in each subregion is the sum of the costs prorated to each of its points, and (ii) the cost per point is independent of the partition).

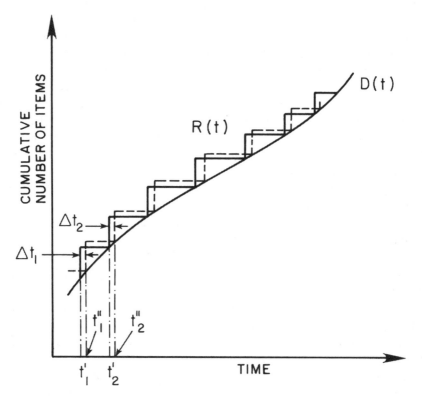

Fig. 4.2   Staggered delivery schedules and their effect on accumulation

The following strategy cuts inventories at the origin by a factor P , while preserving virtually unchanged the motion and holding costs at the destination:

    (i)   produce the items for any shipment in order of production subregion: $\mathbf{P}_1$ first, then $\mathbf{P}_2$, etc,

(ii) On completing production for a subregion, $\mathbf{P}_p$, dispatch the vehicles to the subregion on VRP routes constructed for the subregion alone.

As a practical matter, P does not need to be very large; once it reaches a moderate value (say P $\approx$ 5) additional increases yield decreasingly small benefits. In fact, even if the demand was perfectly constant, it is unlikely that one would choose a P much larger than 5 because larger P's imply shorter production runs within each $\mathbf{P}_p$, which hinders our ability to sequence the production to meet other objectives, such as operating with smoothing worker loads and materials requirements; see Burns and Daganzo (1987).

If production schedules are staggered as described, then the search for the optimal dispatching times should recognize that holding costs will be lower. The analysis could be repeated with a changed holding cost equation (e.g., Eq. (4.8) for the case $c_r \gg c_i$) but this is unnecessary; a suitable (downward) adjustment to the holding cost coefficient, either c_i or c_r, has the same effect and also preserves our results. (If holding costs at the origin can be neglected, the coefficient should be halved; of course, there is no need to pinpoint its value very precisely, since the solution to our problem is robust to errors in the cost coefficients.)

4.4   Identical Customers; Vehicle Loads Not Given

In every case discussed so far, the total cost expression (e.g., Eq. (4.10)) decreases with the vehicle load carried, v_{max} . This should not be surprising, since the larger v_{max} the smaller the total number of vehicles that need to be dispatched. In any practical situation, thus, one would be well advised to use vehicles as large as possible; in fact, as large as the (highway, railway ...) network would allow. Our analysis, however, ignored pipeline inventory cost and did not consider possible route length restrictions. With either one of these complications, it may not always be desirable (or possible) to dispatch full vehicles all the time; vehicle load size becomes a decision variable. We will discuss route length restrictions first, and will then incorporate pipeline inventory into the models. It will be shown that pipeline inventory cost can be ignored for freight that is neither perishable nor extremely valuable, and that it cannot be ignored for passengers. Were it not for this complication, the results of Section 4.3 could be used for one-to-many passenger logistics (e.g., to design a commuter rail network serving a CBD). The section concludes with a discussion of restrictions on the delivery lot size.

4.4.1 Limits to Route Length

If the optimization of Section 4.3 results in very small delivery lot sizes, each vehicle may have to make an unreasonably large number of stops. Very long routes may not be feasible if there are restrictions to the duration of a vehicle tour, due, for example, to labor regulations--see Langevin and Soumis (1989) and Kiesling (1995) for some case studies. This subsection, which may be skipped without loss of continuity on a first reading, explores the consequences of such restrictions.

Tour duration limitations essentially impose a location-dependent limit on the number of stops. Presumably, locations distant from the depot will need to be served with fewer stops than those which are nearer since more time is needed to reach their general vicinity. To recognize this dependence, we use $C_{max}(\mathbf{x})$ for the maximum number of stops around $\mathbf{x}$; we assume that $C_{max}(\mathbf{x})$ varies slowly with $\mathbf{x}$.

Assume first that N is large, so that most delivery districts do not reach all the way to the depot. Then, to minimize distance one should still attempt to design delivery districts of width $(6/\delta(\mathbf{x}))^{1/2}$, while making them long enough to include a desired number of stops at (or near) coordinate $\mathbf{x}$, $n_s(\mathbf{x}) \leq C_{max}(\mathbf{x})$. This yields: length $= n_s(\mathbf{x})/(6\delta(\mathbf{x}))^{1/2}$. The total distance is then given by expressions similar to (4.3) and (4.5a); i.e.:

$$\begin{pmatrix} \text{Total} \\ \text{distance} \end{pmatrix} \approx \left[\sum_i \frac{2 r_i}{n_{s,i}} + k\, \delta^{-1/2}(x_i) \right]$$

$$\approx 2\, N\, E\left(\frac{r}{n_s} \right) + k\, N\, E\big(\delta^{-1/2} \big), \tag{4.12}$$

where $n_{s,i}$ denotes the number of stops per tour used for tours near x_i ; if $n_s(\mathbf{x}) = C$, Eq. (4.12) coincides precisely with these expressions. Although the line-haul distance component (the first term) is somewhat different if $n_s(\mathbf{x})$ varies with $\mathbf{x}$, *the local component remains unchanged.*

Because this expression decreases with $n_{s,i}$, the number of stops per tour should be made as large as practicable. For our problem, the number of stops used near location $\mathbf{x}$ on the ℓth dispatch, $n_s^{\ell}(\mathbf{x})$, should satisfy:

$$n_s^{\ell}(\mathbf{x}) = min\{ C_{max}(\mathbf{x});\, v_{max}/v_{\ell} \},$$

where v_{ℓ} denotes the delivery lot size used for period ℓ. The expression indicates that the vehicle either reaches its route length constraint, or else is filled to capacity.

With this restriction some of the tours may carry less than a full load. As a result, it may appear that neither the total number of vehicle tours nor the line-haul transportation cost (formerly the third term of Eq. (4.7)) are fixed. Would they depend on the specific headways? The short derivation below shows that, while not fixed, the number of tours (and thus the sum of the line-haul and stop costs) can sometimes be approximated by an expression that only depends on the number of headways L ; then, the scheduling and routing decisions can still be decomposed.

Approximation for the number of tours: Assume that **R** can be partitioned into just a few subregions, $\mathbf{P}_p$, with the same limitation on the number of stops: $n_s(\mathbf{x}) \leq C_{max}(\mathbf{x}) \approx C_p$. Characterize each subregion by the number of destinations N_p, and their average distance to the depot $E(r_p)$. We show below, that the number of tours in each subregion only depends on L. As a result, an expression for the total number of tours is developed. The number of tours in period ℓ for subregion p is:

$$\{No.\,tours;\,\ell,p\} = \max\left\{\frac{N_p v_\ell}{v_{max}}; \frac{N_p}{C_p}\right\},$$

and for all periods:

$$\{No.\,tours;\,p\} = N_p \sum_{\ell=1}^{L} \max\left\{\frac{v_\ell}{v_{max}}; \frac{1}{C_p}\right\}$$

$$\geq N_p \max\left\{\sum_{\ell=1}^{L}\frac{v_\ell}{v_{max}}; \frac{L}{C_p}\right\}. \qquad (4.13a)$$

This inequality is a good approximation for the number of tours if rent costs dominate, as then the delivery lot size should be independent of ℓ . The approximation will also be good, for the same reason, if the demand is nearly stationary. Then, we can write:

$$\{No.\,tours;\,p\} \cong N_p \max\left\{\frac{D(t_{max})}{v_{max}}; \frac{L}{C_p}\right\}$$

$$= N_p\left\{\frac{D(t_{max})}{v_{max}} + \max\left[0, \frac{L - L_p}{C_p}\right]\right\} \qquad (4.13b)$$

where $L_p = C_p D(t_{max})/v_{max}$. (This constant represents a critical number of dispatching periods for subregion p . If $L > L_p$, then the lot sizes are so small that the vehicle cannot be filled in subregion p ; the number of stops constraint is binding.) If (4.13b) is a good approximation for the number of tours used in P_p , then the sum of the origin stop cost plus the line-haul cost for all tours (first and third terms of (4.7)) is:

$$
\begin{pmatrix} \text{origin stop} \\ \text{plus line} \\ \text{hault trans -} \\ \text{portation} \\ \text{costs} \end{pmatrix} = \sum_{p=1}^{P} \{No.tours; \, p\}\{c_s + 2c_d \, E(r_p)\}
$$

$$
= \frac{D(t_{max})}{v_{max}} N \left[c_s + 2c_d \, E(r) \right]
$$

$$
+ \sum_{p=1}^{P} \{c_s + 2c_d \, E(r_p)\} \left\{ N_p \max \left\{ 0, \frac{L - L_p}{C_p} \right\} \right\} \qquad (4.14)
$$

which only depends on the dispatching times through L . ∎

For small L the expression is constant, and matches the corresponding components of (4.7), but once L exceeds some of the L_p (some tours hit the length constraint and are only partially filled), it increases with L at an increasing rate.

    The optimal L can be found still as a trade-off between inventory cost, Eq. (4.8), and transportation cost, Eq. (4.7) with two terms revised by Eq. (4.14). Because the revised (4.7) is piecewise linear and convex, the sum of (4.7) and (4.8) has only one local/global minima. The revised derivative of (4.7) with respect to L is now a step function:

$$
\left[c_s + c_d \, k \, E \left(\delta^{-1/2} \right) \right] N + \sum_{L_p < L} \frac{N_p}{C_p} \{2c_d \, E(r_p) + c_s \}, \qquad (4.15)
$$

where the summation only includes p's for which $L_p < L$. The second term represents the cost increase for the extra tours that need to be sent because (some) vehicles cannot be filled to capacity. The first term is the original derivative of Eq. (4.7). In the special case where C_p is the same (C_{max}) for all points, there is only one subregion, with $L_1 = C_{max}D(t_{max})/v_{max}$ and $N_1 = N$. Therefore, the second term is zero if $L \le C_{max}D(t_{max})/v_{max}$, and equals $(N/C_{max})(2c_dE(r) + c_s)$ otherwise. The optimal L can be found as follows: If there is a value of L for which the sum of (4.15) and the derivative of (4.8) equals zero, then that value is optimal; otherwise, the optimal value is the L_p for which the sum changes sign.

Because (4.15) is larger than before, the optimal L will tend to be smaller and the resulting cost greater. This is intuitive; with limits to route length it may be advisable to increase the lot sizes (by reducing L) to make sure that most of the vehicles travel full.

Our results assume that all customers share the same L and v_ℓ . Although this simplification facilitates production scheduling, it may also increase logistics costs when C_p changes significantly across subregions. If a different L can be used for different subregions, then fewer dispatching intervals and larger delivery lot sizes can be used for subregions with a low C_p; all the vehicles can be filled as a result. A strategy (a set of dispatching times and delivery districts) can then be tailored to each one of the subregions independently of the others. We will explore this point – the determination of routing/dispatching strategies that vary in time and space – more thoroughly in the next subsection.

To conclude our discussion on route length restrictions, we must consider the case with few vehicle tours, $N << C^2$. But this is very simple. We already stated that, for this case, the transportation cost is insensitive to the number of stops per vehicle. Hence, route length restrictions do not influence either the optimal dispatching strategy or the final cost.

4.4.2 Accounting for Pipeline Inventory Cost

In all the optimization problems described so far we have found a solution which minimizes the sum of the motion cost, the holding (rent) cost and the stationary inventory cost. We did not consider the pipeline inventory cost of the items in the vehicles.

It had been pointed out in Chapter 2 that the pipeline inventory cost per item was $c_i t_m$, where t_m is the average time an item spends inside a vehicle. On average an item spends in a vehicle a time approximately equal to one-half of the duration of the tour. If the vehicle travels at a speed s , and takes t_s time units per stop, the duration of a tour with n_s stops and d distance units long is $d/s + (n_s + 1)t_s$; thus:

$$t_m \approx \frac{1}{2}\left[\frac{d}{s} + (n_s + 1)t_s\right],$$

and

$$c_i t_m \approx \frac{1}{2} c_i \left(\frac{d}{s} \right) + \frac{1}{2} c_i t_s \left(n_s + 1 \right). \qquad (4.16)$$

Added for all items for all L shipping periods, the pipeline inventory cost becomes a simple function of the total number of (item-miles), (items) and (item-stops):

$$\begin{pmatrix} \text{Total} \\ \text{combined} \\ \text{pipeline} \\ \text{cost} \end{pmatrix} \approx \frac{c_i}{s} \{ \text{No. item - miles} \} + \frac{c_i t_s}{2} \{ \text{No. items} \}$$

$$+ c_i t_s \{ \text{no. of item - stops} \}.$$

The total number of items is $D(t_{max})N$. The total number of item-miles and item-stops can be obtained easily if there are no route length restrictions. In that case vehicles travel full (from the depot) and every stop delays on average $v_{max}/2$ items; therefore, the total number of item-stops is NL $v_{max}/2$. Similarly, each vehicle carries on average $v_{max}/2$ items and the item-miles equal the product of the vehicle-miles and ($v_{max}/2$). We have already seen in the derivation of (4.7) that the total vehicle-miles are:

$$\begin{pmatrix} \text{Total} \\ \text{combined} \\ \text{distance} \end{pmatrix} \approx \frac{2(r)D(t_{max})N}{v_{max}} + kNLE\left(\delta^{-1/2} \right).$$

Thus, the pipeline inventory cost can also be expressed as a function of the decision variables through L alone:

$$\left[\frac{c_i E(r)D(t_{max})N}{s} \right] + \left[\frac{c_i kN}{2s} E\left(\delta^{-1/2} \right) v_{max} \right] L$$

$$+ \frac{c_i t_s}{2} \left[D(t_{max})N + N v_{max} L \right]. \qquad (4.17)$$

As a function of L , this expression is similar to Eq. (4.7), but it increases much more slowly: at a rate $[Nc_i v_{max}/2]\{t_s + kE(\delta^{-1/2})/s\}$ as opposed to $N\{c_s$

+ $c_d k E(\delta^{-1/2})\}$. Normally, the quantity $c_i v_{max} t_s$ representing the cost of delay to the items in a full vehicle during a stop should be several orders of magnitude smaller than c_s (the truck cost and driver wages during the stop). Likewise, the quantity $c_i v_{max} /s$ representing the inventory cost of a full truck per unit distance should be much smaller than c_d (the vehicle operating cost per unit distance, including driver wages). Thus, if pipeline inventory costs had been considered from the beginning, the results would not have changed.

If the items are so expensive that the pipeline inventory component cannot be neglected, then Eq. (4.17), unlike (4.7), increases with v_{max}. One could thus imagine a situation where a v_{max} smaller than the maximum possible might be advantageous; the vehicle loads cannot be assumed to be known. The transportation of people is a case in point, where the inventory cost of the items carried (the passengers) vastly exceeds the operating cost. That is why airport limousine services do not distribute people from an airport to the hotels in the outlying suburbs in large buses; this would result in unacceptably large routes, with some passengers spending too much time in the vehicle (see Banks, et al. 1982, for a discussion). Let us now see how to select the routes and schedules for a system carrying items so valuable that vehicle loads are not necessarily maximal.

Without an exogenous vehicle load, the total transportation cost no longer can be expressed as a function of L alone, as in Eq. (4.7); the total vehicle-miles and the number of tours depend on the specific vehicle-loads used, and this has to be recognized in the optimization.

To cope with this complication, we will consider a set of strategies more general than the ones just examined, but will analyze them less accurately. We will now allow different parts of **R** to be served with different delivery headways at the same time. To do this, we define the smooth and slow varying function $H(t, \mathbf{x})$, which represents the headways one would like to use for destinations near $\mathbf{x}$ at times close to t.

(Until now we had assumed that the headways were only a function of t: $H(t, \mathbf{x}) = H(t)$. As a result, the optimal dispatching times $\{t_\ell\}$ could be found with the exact numerical techniques of Chapter 3; or if D(t) was slow varying, with the CA approach, as described in that chapter.)

For the present analysis we also seek a function $n_s(t, \mathbf{x})$ which indicates the number of stops made by tours near $\mathbf{x}$ at a time close to t. Of course, this number cannot be so great that the vehicle capacity is exceeded; the following must be satisfied:

$$\{n_s(t, \mathbf{x}) D'(t)\} H(t, \mathbf{x}) \le v_{max}. \tag{4.18}$$

The quantity in braces represents the combined demand rate at the n_s destinations visited by a tour, and the left side of the inequality the load size carried by the vehicle.

(The approach we had used assumed that (4.18) was a pure equality, so that n_s was only a function of t, $n_s(t) = v_{max}/[H(t)D'(t)]$, implicitly given by H(t) .)

Like H(t, x) , the function $n_s(t, x)$ will be allowed to be continuous and slow-varying during the optimization. Once H(t, x) and $n_s(t, x)$ have been identified, a set of delivery districts and dispatching times consistent with these functions must be found. This will be illustrated after the optimization has been described.

Let us write the total logistic cost per item that items at time-space point (t, x) would have to pay if the parameters of the problem were the same at all other times and locations, i.e.,

$$D'(t) = D',$$
$$\delta(x) = \delta, \text{ and}$$
$$r(x) = r.$$

As explained in Chapter 3, the decision variables, H and n_s , that minimize such an objective function will become the sought solution, varying continuously with t and x [H(t, x) and $n_s(t, x)$] . The minimum value of the objective function for these coordinates z(t, x), is the CA cost estimate.

Noticing that a vehicle load consists of $D'n_sH$ items and a delivery lot of D'H items, we can express the total motion cost per item as:

$$z_m = \frac{2rc_d}{(D'n_sH)} + c_d \, k \, \delta^{-1/2} \, \frac{1}{(D'H)} + c_s \left(\frac{1}{D'H}\right)$$
$$+ c_s \frac{1}{(D'n_sH)} + c'_s .$$
$$(4.19a)$$

This expression, consistent with Eqs. (2.5e) and (4.7) when the number of tours is large compared with the number of stops per tour, has an intuitive physical interpretation. Each tour incurs a cost ($2rc_d + c_s$) for overcoming the line-haul distance and stopping at the origin, which prorated to all the items in the vehicle yields the first and fourth terms of (4.19a). The tour also incurs a cost ($c_d k\delta^{-1/2} + c_s$) for each local stop and detour, which prorated to the items in a delivery lot, yields the second and third terms of the expression. The last term is the (constant) cost of handling each item.

Thus, the first two terms are the cost of overcoming line-haul and local distance (assuming that many tours are needed) the third term is the cost of stopping at the destinations; the fourth the cost of stopping at the origin, and the last one the handling/loading cost.

The holding costs can be expressed in a similar manner. For the pipeline inventory cost per item, z_p, we use (4.16) and the distance per tour (4.1) with $C = n_s$:

$$z_p = c_i \left(\frac{r}{s}\right) + c_i \, k \, \delta^{-1/2} \frac{n_s}{2s} + c_i \frac{t_s}{2} n_s + \frac{1}{2} c_i t_s . \qquad (4.19b)$$

As with (4.19a), the four terms correspond to times spent in line-haul travel, local travel, destination stops, and at the origin. The stationary inventory cost per item averages

$$z_s = c_i H \qquad (4.19c)$$

if we count it both at the origin and the destination. The rent cost can be ignored because if items are expensive compared to transportation costs, they will certainly satisfy $c_i \gg c_r$; thus $c_h = (c_i + c_r) \approx c_i$, and we can write:

$$z_s = c_h H. \qquad (4.19d)$$

(Inclusion of rent costs would pose a problem because rent does not depend only on local characteristics such as H and n_s . An exception arises if the demand is stationary in time, $D'(t) = D'$, because then the optimal solution is also stationary; i.e., $H(t, \mathbf{x})$ is independent of t, and the rent cost is $c_r H$.)

If instead of H (and as is often done in the literature) we use the delivery lot size $v = D'H$ as a decision variable, keeping n_s as the other variable, then the sum of Eqs. (19) can be expressed as:

$$z = \alpha_0 + \alpha_1 \left(\frac{1}{n_s v}\right) + \alpha_2 \left(\frac{1}{v}\right) + \alpha_3 (n_s) + \alpha_4 (v), \qquad (4.20a)$$

where the α_0 , ... , α_4 are the following interpretable cost constants, which will be used from now on:

$\alpha_0 = (c'_s + c_i r/s + c_i t_s/2)$; handling and fixed pipeline inventory cost per item,

$\alpha_1 = (2rc_d + c_s)$; transportation cost per dispatch,

$\alpha_2 = (c_d k\delta^{-1/2} + c_s)$; transportation cost added by a customer detour,

$\alpha_3 = 1/2 \ c_i(k\delta^{-1/2}/s + t_s)$; pipeline inventory cost per item caused by a customer detour and the ensuing stop,

$\alpha_4 = c_h/D'$; stationary holding cost of holding one item during the time $(1/D')$ between demands.

With the new notation, (4.18) becomes:

$$n_s v \le v_{max} ; \qquad (4.20b)$$

in addition, we require

$$n_s \ge 1 \qquad (4.20c)$$

Equation (4.20a) is a "logistic cost function" (LCF) that relates the cost per item distributed to the decision variables of our problem. We will see in the remainder of this book that the determination of a realistic LCF is perhaps the most important step in the design of a logistics system with the approach espoused in this monograph. In the present case, the minimum of Eq. (4.20a) subject to these inequalities is the solution to our problem. Note that α_0 can (and often will) be omitted for optimization purposes. Note as well that, with a small modification to the expressions for α_1 , α_2 , and α_3 , Eq. (4.20a) also applies to the VRP case with a small N (compared with n_s^2); k should be replaced by k' and the term $2rc_d$ should be omitted. We will assume for the remainder of this section that the α_1, α_2, and α_3 for large N are used in the optimization; if the resulting n_s found in an application is inconsistent with these values, then the α's should be changed to recognize that N is "small". Our qualitative discussion also applies to this case, which is very similar.

   We now identify a condition under which the pipeline inventory term $(\alpha_3 n_s)$ can be neglected, and show that in that case (4.20b) should be a pure equality. This has been pointed out in Daganzo (1985a) and Burns et al. (1985).

The full vehicle condition: For any integer, n_s , a feasible solution to Eqs. (20) is $v = v_{max}/n_s$, which (ignoring α_0) yields:

$$z(n_s) = \frac{\alpha_1}{v_{max}} + \frac{\alpha_2 n_s}{v_{max}} + \alpha_3 n_s + \alpha_4 \frac{v_{max}}{n_s}.$$

An upper bound, z^u , to the minimum of Eqs. (4.20), z^* , is obtained from $z(n_s)$, using $n_s = 1$ if $v_{max}(\alpha_4/\alpha_2)^{1/2} < 1$, and $n_s \approx v_{max}(\alpha_4/\alpha_2)^{1/2}$ otherwise; that is:

$$z^* \le z^u \approx \frac{\alpha_1}{v_{max}} + 2\left((\alpha_2 \alpha_4)^{1/2} + (\alpha_3 v_{max})\left(\frac{\alpha_4}{\alpha_2}\right)^{1/2}\right) \quad , if \; v_{max}\left(\frac{\alpha_4}{\alpha_2}\right)^{1/2} \ge 1$$

$$\approx \frac{\alpha_1}{v_{max}} + \frac{\alpha_2}{v_{max}} + \alpha_4 v_{max} + \alpha_3 \qquad , otherwise.$$

A lower bound to the optimal cost is obtained by neglecting the pipeline inventory term $\alpha_3 n_s$ of Eq. (4.20a), and optimizing (4.20). We see at a glance that (4.20a) decreases with n_s for any v ; thus, one will always choose the largest n_s satisfying (4.20b): $n_s \approx v_{max}/v$. (Note that if $v < v_{max}$, then (4.20c) holds.)

If this value is substituted for n_s in (4.20a), without its first and fourth terms, we obtain a function $z(v)$,

$$z(v) = \frac{\alpha_1}{v_{max}} + \frac{\alpha_2}{v} + \alpha_4 v,$$

whose minimum (subject to $v \le v_{max}$) is a lower bound, z^ℓ . Its expression – see the constrained EOQ trade-off reviewed in Chapter 3 – is:

$$z^* \ge z^\ell \approx \frac{\alpha_1}{v_{max}} + 2\left(\alpha_2 \alpha_4^{1/2}\right) \quad , if \left(\frac{\alpha_2}{\alpha_4}\right)^{1/2} \le v_{max}$$

$$\approx \frac{\alpha_1}{v_{max}} + \frac{\alpha_2}{v_{max}} + \alpha_4 v_{max} \quad , otherwise. \tag{4.21}$$

Notice that the expressions for z^u and z^ℓ are almost identical: $z^u - z^\ell = \alpha_3 v_{max} (\alpha_4/\alpha_2)^{1/2}$ if $v_{max}(\alpha_4/\alpha_2)^{1/2} \ge 1$, and $z^u - z^\ell = \alpha_3$, otherwise.

The relative difference between any two of z^u, z^* and z^ℓ should be lower than ε, the ratio of the maximum value of $(z^u - z^\ell)$ to the second term in the upper part of (4.21), which bounds z^ℓ from below; i.e.: $\varepsilon = (\alpha_3 v_{max}/2\alpha_2)$. The numerator of this constant, $\alpha_3 v_{max}$, is the pipeline inventory cost accruing to a full vehicle for one delivery detour; the denominator is double the vehicle motion cost per detour. For most commodities this ratio is orders of magnitude smaller than 1, so that the lower and upper bounds will nearly coincide.

In summary, if $\varepsilon \ll 1$, then filling the vehicles (as done with the strategy leading to z^u) is near optimal; the resulting cost is close to the lower bound, obtained without pipeline inventory costs. ■

The incentive to fill vehicles, used so far in this chapter, does not apply if $\varepsilon = \alpha_3 v_{max}/2\alpha_2$ is large compared with 1. The minimization problem described by Eqs. (4.20) then yields a strict inequality for (4.20b). We now examine the solution to this minimization problem with varying conditions in time-space (Daganzo and Newell, 1985).

The unconstrained minimum of (4.20a) can be obtained numerically, and it can also be expressed analytically as a function of one single parameter β. To see this, let n_s be close to the unconstrained minimum of (4.20a): $n_s \approx (\alpha_1/\alpha_3 v)^{1/2}$; then $z^*(v) = 2(\alpha_1\alpha_3/v)^{1/2} + \alpha_2/v + \alpha_4 v$. This expression reflects an achievable cost if $n_s > 1$. Because $z^*(v)$ is convex, its minimum is the root of $dz^*(v)/dv = 0$. Using $v' = (\alpha_1\alpha_3 v)^{1/2}/\alpha_2$, we can express this equation in terms of v' as follows:

$$\beta(v')^4 = 1 + v'; \quad i.e., \quad \beta = (v')^{-4} + (v')^{-3}$$

where $\beta = \alpha_4\alpha_2^3/(\alpha_1\alpha_3)^2$. When v' is small compared with 1 the second term in the last expression can be neglected; in this case the solution is: $v' \approx \beta^{-1/4} \ll 1$ for $\beta \gg 1$. Conversely, if v' is large compared with 1, i.e., $\beta \ll 1$, the first term can be neglected and the solution becomes $v' \approx \beta^{-1/3}$. The largest of the two extreme solutions can be used as a rough approximation when $\beta \approx 1$. The optimal vehicle load is $n_s v = v'\alpha_2/\alpha_3$, and $n_s = \alpha_1/\alpha_2 v'$. If the vehicle load is smaller than v_{max} and $n_s > 1$, then the solution can be accepted. (This happens if $\alpha_2 v' < \alpha_1$ and $\alpha_3 v_{max}$). The optimal H and z can also be expressed as a function of v', and thus of β.

Without pipeline inventory, the solution z^* is as Eq. (4.21). In that case we see that $z^* \approx z^\ell$ increases linearly with α_1 (which also increases linearly with the distance from the depot r). Because there is an intercept, both z^* and the total cost per unit time $ND'z^*$ increase "less-than-proportionately" with r ; the ratio of cost to distance decreases. We also see that z^* decreases with the demand rate per customer D' , but increases with the spatial density of customers δ if their aggregate demand rate ND' (i.e., δD') is constant. However, the total cost per unit time ND'z* is non-decreasing with D' . While not so obvious, these scale economies are also shared by the solution to Eqs. (4.20) as just described. While ND'z* increases with D', z* decreases; the optimal cost also increases less than proportionately with distance from the depot.

To estimate cost for a problem with varying D'(t) , $\delta(\mathbf{x})$ and r($\mathbf{x}$) , one would need to average the analytical solution over t and $\mathbf{x}$. Although it may be possible to do this in closed form using statistical approximation formulas for expectations (these indicate that cost increases with variable conditions; see Problem 4.3), a few numerical calculations should suffice. One could calculate z* for all the D(t_{max})N items demanded, using their respective t and $\mathbf{x}$, but this would be too laborious. Instead, one can partition the time axis into m = 1, ... , M intervals and $\mathbf{R}$ into p = 1 , ... P subregions so that each (m, p) combination includes roughly the same amount of demand. We use any interior point (t , $\mathbf{x}$) of each combination to calculate both the parameters of the optimization and the resulting cost, z^{mp}. The estimated cost is then the arithmetic average of the z^{mp}. Section 4.5 shows how a detailed solution can be developed.

4.4.3 Storage Restrictions

We conclude our analysis of systems with identical customers with a discussion of storage restrictions. This subsection can be skipped without loss of continuity.

Over the short term, it may not be possible to change the amount of storage at the destinations, and one may wish to design the distribution operation recognizing this restriction. This problem, arising with the distribution of gasoline to service stations (Brown and Graves, 1980, Dror et al., 1985, and Webb, 1989) is not difficult to model.

Let $v°(\mathbf{x})$ denote the maximum allowable accumulation of items (i.e., the available storage space) for destinations at or near $\mathbf{x}$, assumed to be smaller than v_{max}. This function should be slow varying and independent of t. Then, the total rent cost over the study period is approximately:

$$Total\ rent\ cost = c_r t_{max} \int_R \delta(x) v°(x) dx,$$

which is independent of the decision variables H and n_s ; (or n_s and v) .

For any time and location, the optimal variables and cost are still the minimum of (4.20a) with two minor differences: (i) the holding cost parameter of α_4, c_h, should not include rent costs ($c_h \approx c_i$) since they are fixed in the short term; and (ii) in addition to (4.20b) and (4.20c), one must include a constraint representing the storage restrictions at the destinations:

$$v \le v^\circ. \tag{4.20d}$$

The minimum of Eqs. (4.20) is the solution to the problem.

If items are cheap, so that the pipeline inventory can be neglected, then inspection of (4.20a, b, c, and d) reveals that for any feasible v, one will choose the largest possible n_s satisfying (4.20b). We can thus restrict ourselves to values of n_s and v satisfying (4.20b) exactly, and solve instead:

$$\frac{\alpha_1}{v_{max}} + \min\left\{ \frac{\alpha_2}{v} + \alpha_4 v \ , \quad with \quad v \le v^\circ \ and \ v_{max}/v = 1,2,3...\right\};$$

or in terms of $n_s = v_{max}/v$:

$$\frac{\alpha_1}{v_{max}} + \min\left\{ n_s \frac{\alpha_2}{v_{max}} + \frac{\alpha_4 v_{max}}{n_s} \ , \quad with \quad n_s = 1,2,... \ge v_{max}/v^\circ \right\}.$$

This is an integer constrained EOQ. Since $v_{max} > v^\circ$, then $n_s^* > 1$ and the above minimum can be approximated (as discussed in Example 2 of Section 3.5) by ignoring the integrality requirement – Equation (3.5b) of Chapter 3 gives the result with a different notation: $A \equiv \alpha_4$; $B \equiv \alpha_2$; and $v_{max} \equiv v^\circ$. We have ignored the possibility $v_{max} \le v^\circ$, because if each destination can hold a full vehicle load, then the storage restriction plays no role; i.e., the constraint to the number of stops is: $n_s = 1, 2, ...$, independent of v° .

If items are expensive, the solution can be obtained numerically for various t and $\mathbf{x}$. In both cases, the average of z^* over $(t, \mathbf{x})$ can be used to estimate the total cost with variable conditions.

Because the estimated total cost depends on $v^\circ(\mathbf{x})$, this method can be used to explore the attractiveness of changes in the amount of storage space perhaps being contemplated as a strategic long run decision. One would have to balance the changes in rent cost against the changes in the remaining costs just discussed.

In most cases, it is likely that either $c_r \ll c_i$ or $c_r \gg c_i$ (either rent costs or inventory costs dominate holding costs) and then the optimal amount of storage space easily follows from the results already presented. We have argued already that if $c_r \ll c_i$, rent costs should not influence the routing/dispatching strategy. For this case, then, each destination should be designed to have barely enough space to hold the maximum delivery lot sizes it will receive with the just discussed optimal strategy: $v^{\circ*}(\mathbf{x}) = \max_t \{v^*(t, \mathbf{x})\}$. (This should certainly be the case for passenger terminals such as transit stations.) If, on the other hand, $c_r \gg c_i$, then rent costs will influence dispatching. Fortunately, with dominant rent costs, pipeline inventory can be neglected and vehicles should travel full; as a result, the decomposition technique illustrated at the beginning of Section 4.3 can be used to separate dispatching and routing decisions, yielding the (constant) delivery lot size. Again, the amount of storage space at the destination should be just enough to accommodate this lot size.

4.5 Implementation Considerations

This section describes how specific solutions can be designed from the optimization results in prior sections. It also discusses systematic ways for fine-tuning the designs.

We saw in Chapter 2 that changes in the input parameters of an EOQ optimization have a dampened effect on the decision variables; this is also true for the objective function now at hand. Thus, if $D(t)$ and $\delta(\mathbf{x})$ change slowly, the decision variables H (or v) and n_s will change even more sluggishly over t and $\mathbf{R}$. Because, as with the EOQ optimization, the decision variables themselves do not need to be set very precisely, it should be possible to identify large regions of the time-space domain where the decision variables can be set constant without a serious penalty.

For our problem with identical customers, the partition is easily developed: (i) divide the time axis into m = 1, 2, ... , M periods with nearly constant demand rates; and (ii) partition $\mathbf{R}$ into p = 1, 2, ... P subregions with similar customer density and distance to the depot. The subregions and time periods should be large enough to include respectively several delivery districts and several headways. This ensures that the number of stops in each district can be close to ideal, and that the theoretical headway $H(t, \mathbf{x})$ can be approximated with an integer number of dispatches. We anticipate now that, by designing a different spatial partition for every time period, this method can be extended to situations with different customers and time varying customer densities.

4.5.1   Clarens and Hurdle's Case Study

An application of the technique for a very similar problem has been reported by Clarens and Hurdle (1975). These authors explored the best way of laying out transit routes from a CBD to its outlying suburbs. They assumed that the demand was stationary and changed with position.

They describe the solution in terms of slightly different variables and inputs, but the differences are only superficial. They define the vehicle operating cost as a function of time (and not distance), c_t , and do not explicitly account for the number of stops; instead they assume that one knows from empirical observations the time that it takes for a bus to cover one unit area – a constant, $\tau(\mathbf{x})$, that can vary with position. They define the demand as a density per unit area and unit time, $\lambda(\mathbf{x})$, which changes with position. Instead of a distance from the CBD, $r(\mathbf{x})$, they define an express (line-haul) travel time, $T(\mathbf{x})$, and as a decision variable they use the area of a bus service zone, $A(\mathbf{x})$, instead of $n_s(\mathbf{x})$. Thus, they work with the following logistic cost function, which is equivalent to (4.20a):

$$z = \frac{2c_t T}{A\lambda H} + \frac{\tau c_t}{\lambda H} + c_i\{T + \tau\, A/2\} + c_h H/2, \qquad (4.22)$$

where the bus load, $A\lambda H$ is restricted to be below $v_{max} = 45$ passengers. Note that the mathematical optimization problem is analogous to Eqs. (4.20).

Figure 4.3, displays the demand distribution for the case study presented in that reference. Figure 4.4 depicts the worksheet that was used to design the vehicle routes (the reference does not seem to recognize the benefits of elongating the zones toward the depot) and Table 4.1 compares the actual and ideal zone sizes.

Given the close agreement between these two columns of figures and the robustness of the CA solution to small departures from the recommended settings, one would expect to have a cost that is very close to the minimum.

The Clarens-Hurdle case study is to this author's knowledge the only published example where the CA guidelines have been translated into a proposed design for a two-dimensional problem.

On reviewing the procedure, it becomes clear that a great deal of human intuition is required to complete a design. Furthermore, careful efforts notwithstanding, the designer may miss opportunities for small improvements at the margin that depend on specific details (e.g., stop locations, street intersections, etc.) of the particular problem. It might be worthwhile to use fine-tuning software to find these possible improvements if any exist.

4.5.2 Fine-Tuning Possibilities

The rest of this section describes the results of some experiments where fine-tuning software was used to improve detailed VRP solutions developed quickly from the guidelines given in Section 4.2. The discussion is based on Robusté et al. (1990).

These authors tested simulated annealing (SA) as a technique that is well suited for fine-tuning purposes. The brief discussion of simulated annealing provided in this reference is included as Appendix B. The technique is attractive because:

(i)   A prototype computer program can be developed quickly for most problems since the SA logic is very simple. (These authors developed software for the VRP, from scratch, in about three man-days.)

(ii) The optimization can be controlled by means of input variables (called initial "temperature" and "cooling rate" or "annealing schedule") which determine how much the algorithm is allowed to increase (worsen) the objective function at different stages of the process in the hope of finding larger reductions later.

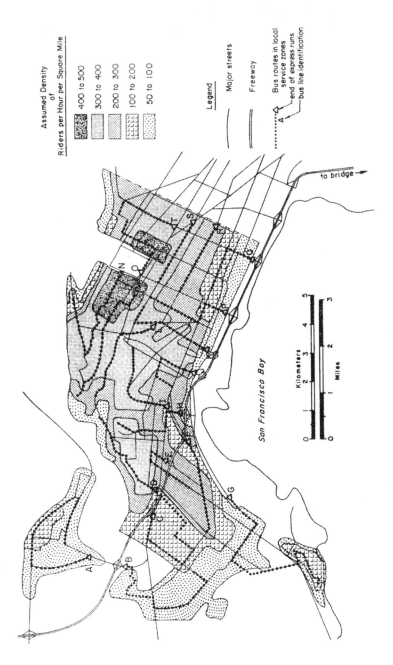

Fig. 4.3   Demand distribution for a transit line design problem.
(Source: Clarens and Hurdle, 1975).

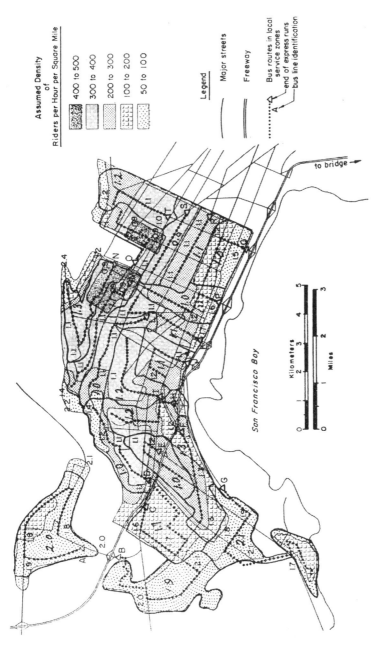

Fig. 4.4   Worksheet for a transit design problem
(Source: Clarens and Hurdle, 1975)

Table 4.1   Results of the transit line design process
(Source: Clarens and Hurdle, 1975)

Zone	Area (square miles)		Headway (minutes)	Average Load On Bus (persons)	Load Factor	$T(x,y)$ (minutes)	$r(x,y)$ (min./sq mi)
	Actual	$A^*(x,y)$					
A	2.0	1.9	13	35	78	27	9
B	1.9	1.9	14	31	69	27	10
C	1.7	1.3	10	43	96	25	10
D	1.0	1.1	9	39	87	26	11
E	1.0	1.2	11	40	89	24	9
F	1.3	1.4	14	36	80	26	10
G	2.1	1.9	8	27	60	21	9
H	1.2	1.5	7.3	38	84	22	8
I [a]	1.2	1.2	7	45	Full	26	8
J	1.1	1.2	7	36	80	20	8
K	1.1	1.1	9.0	38	84	19	10
L	1.0	1.0	6.7	45	Full	24	13
M [a]	1.1	0.9	6.7	45	Full	26	13
N [a]	1.3	0.8	5.8	45	Full	29	16
O [a]	0.9	1.0	8	43	96	25	11
P	1.0	1.0	9	30	67	17	11
Q	1.0	1.3	10	23	51	15	8
R	1.1	1.3	8	35	38	20	8
S	0.9	1.0	7	40	89	21	10
T	1.2	1.5	7	39	87	22	7

[a] Zones where $A^*(x,y) = A_e(x,y)$.

Simulated annealing is known to converge in probability to the global optimum of combinatorial optimization problems, such as those arising when designing *in detail* logistics systems. Unfortunately, convergence is slow. To be guaranteed, the initial temperature has to be very large and the cooling rate very slow; the computer time required rapidly becomes prohibitively long with increasing problem size. However, with an overall idea of the system's structure, and a near optimal initial solution as would be obtained with nondetailed methods, the scope of the annealing search can be restricted. As demonstrated in Robusté et al. (1990), a low initial temperature achieves that. It prevents the search from wandering away from the initial solution, while systematically testing variations that exploit the details (specific locations of customers, for example.)

One of the examples in this reference considers a VRP problem with N = 500 points (randomly located according to a uniform density in a 6-inch by 10-inch rectangle), C = 45 stops per tour and a centrally located depot; distances are Euclidean. For this test the VRP formula, Eq. (4.5b) with k ≈ 0.57, predicts a total distance averaging 179 inches. With a high initial temperature, the SA approach yielded tours that were **very** long in reasonable times; after *one day* of computation it obtained a set of tours 180.4 inches long (Figure 4.5). This was reasonable, but longer than the hand constructed tours (Figure 4.6) using the VRP guidelines presented earlier. When the hand constructed tours were used to initiate SA with a low initial temperature, the SA algorithm found enough modifications to reduce the total length by about four percent – to 173.6 inches.

Other tests performed in this reference show that the non-detailed approach, fine-tuned with SA, can obtain solutions with objective functions as low as those currently believed to be optimal. The efficiency of the two-step approach has also been demonstrated in practice – the (non-detailed) results in Burns and Daganzo (1987) were used in conjunction with SA to schedule the assembly lines in some GM plants.

These observations are in agreement with our philosophical conclusions in Chapter 2. Like the evolution processes in nature, to design a complex logistic system it seems best to develop a preliminary design based on the overall characteristics of the problem, and use the details later to fine-tune the preliminary design. This view has been adopted in the recent works of Langevin and StMleux (1992) and Hall et. al. (1994). Although the CA approach and the SA algorithm seem to be ideal companions for this two-step approach, other methods may also be useful. The critical thing is not the specific approach for each step, but the fact that the first step disregards details in searching over all possible solutions, and the second step – restricted to a small subset of possible solutions – incorporates all the details. Perhaps other computer fine-tuning methods will improve on SA (Neural Networks and Tabu Searches are currently in vogue; see Hopfield and Tank, 1985, and Glover, 1989 and 1990, for reviews). But the improvement should not be measured only on computation grounds; the ability to develop the software quickly is just as important.

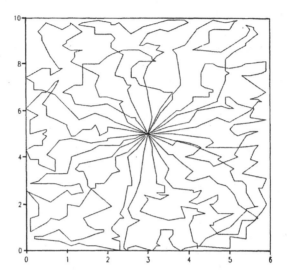

Fig. 4.5   500 point VRP. SA solution with C = 45. 12 tours with total length = 180.4". (Source: Robusté et. al., 1990)

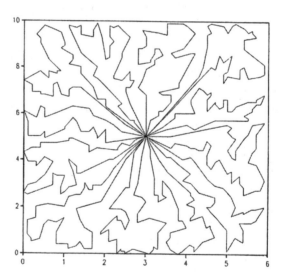

Fig. 4.6   500 Point VRP. C=45. Manual solution. 12 tours with total length = 179.8 inches. (Source: Robusté et. al., 1990)

4.6   Different Customers: Symmetric Strategies

The rest of this chapter considers different customers. Extensions to the strategies we have just described, without exploiting the differences among customers, are easy to develop; they are described in this section. Asymmetric strategies, which allow different customer types to be served differently, are explored in Section 4.7. Conditions under which these more complex strategies are likely to be of benefit are also discussed in that section.

Let us allow $D_n(t)$ to vary across customers, n, and possibly to be non-stationary. With this generalization, even if the demand is stationary, D'_n can vary across n. With many customers the individual demand rates should be treated as "details," which we try to avoid. To this end, an expected demand density rate per unit area is used instead of the specific $D_n(t)$'s. This parameter, $\lambda(t, \mathbf{x})$, is assumed to vary slowly with time and location so that the demand in a subregion, $\mathbf{P}_p$ of $\mathbf{R}$ (large enough to contain several destinations but of small dimensions relative to $\mathbf{R}$) during a time interval $[t_{m-1}, t_m)$ is:

$$\int\limits_{t=t_{m-1}}^{t_m} \int\limits_{\mathbf{x} \in \mathbf{P}_p} \lambda(t, \mathbf{x}) \, d\mathbf{x} \, dt. \tag{4.23a}$$

Similarly, we define a customer density, $\delta(t, \mathbf{x})$, which is also allowed to vary with time. Note that we are allowing here for the number and locations of customers to change with time; all we require is that these changes can be approximated with functions $\delta(t, \mathbf{x})$ and $\lambda(t, \mathbf{x})$, that vary smoothly with t and $\mathbf{x}$.

Demand uncertainty, an important phenomenon when the tours have to be planned before the demand is known at the destinations, will also be considered in this section. It will be captured by an index of dispersion function, as described below.

Take a partition $\{\mathbf{P}_1,...,\mathbf{P}_p,...,\mathbf{P}_P\}$ of $\mathbf{R}$ and a partition of time into consecutive intervals $\tau_m = [t_{m-1}, t_m)$, and let D_{mp} represent the actual number of items demanded in $\mathbf{P}_p$ during τ_m. The parameter $\lambda(t, \mathbf{x})$ can then be defined as the *average* demand rate density, so that (4.23a) now gives the *mean* of D_{mp}. We assume that, for any partition, the variables D_{mp} are independent, and identically distributed. Then their variance can be expressed as:

$$var\{D_{mp}\} \cong \int_{\tau_m} \int_{P_p} \lambda(t,x)\gamma(t,x)\,dx\,dt, \qquad (4.23b)$$

where $\gamma(t, x)$ is an "index of dispersion", with "items" as its physical dimension. A special case of this model arises if each customer's demand fluctuates independently of other customers, either like a stochastic process with independent increments – such as a compound Poisson process or a Brownian motion process. Although in most cases a fixed γ should capture demand fluctuations well, we allow $\gamma(t, x)$ to vary slowly with t and x. An index equal to zero represents known demand; no uncertainty. This case is examined next.

For consistency with the literature, we continue to use $H(t,x)$ and $A(t, x)$ as the decision variables instead of n_s and v. Both formulations are equivalent, since there is a 1:1 correspondence between two sets of variables – the number of stops in a tour is the number of customers in its district, which is given by $n_s \approx \delta(t, x)A(t, x)$, and the delivery lot size is the consumption during a headway in the area around a customer: $v \approx \lambda(t, x)H(t, x)/\delta(t, x)$. Making these substitutions in Eq. (4.20a), and recognizing that $D' = \lambda/\delta$, the cost per item at (t, x) can be expressed as:

$$z = \frac{\alpha_1}{A\lambda H} + \frac{(\delta\alpha_2)}{\lambda H} + (\delta\alpha_3)A + c_h H + \alpha_0, \qquad (4.24a)$$

where α_1, α_2, and α_3 are the constants defined in connection with Eq. (4.20), which now can vary in both time and space; constraints (4.20b), (4.20c), and (4.20d) become:

$$\lambda A H \le v_{max}, \qquad (4.24b)$$

$$\delta A \ge 1 \quad \text{and} \qquad (4.24c)$$

$$\lambda H \le v^{\circ}\delta. \qquad (4.24d)$$

Note that the new logistic cost function has the same functional form with respect to its decision variables as Eqs. (4.20a) and (4.22) have with respect to theirs. Therefore, everything said in connection with these equations and their solution also applies now. The minimum of Eqs. (4.24) for different values of t and x can then be used to construct a near-best symmetric strategy.

The important thing to remember here is not the form of Eqs. (4.24), but the process followed to derive them and use them. This process is quite general and can be used for problems involving various peculiarities. Because it is impossible here to discuss all possible situations, the process is only illustrated with three examples involving stochastic phenomena and requiring some modifications to the equations. The first example (Sec. 4.6.1) arises where items are indivisible and the expected demand per customer per headway is less than one item; the second (Sec 4.6.2) when the customer demands are not known until the vehicles make the stop; and the third (Sec 4.6.3) when the vehicles make coordinated adjustments to their routes as demand information becomes known. These sections can be skipped on a first reading.

4.6.1   Random Demand: Low Customer Demand

Equations (4.24) implicitly assume that each customer is visited each time – the number of stops is equal to δA . But if items are indivisible (as opposed to fluids, or very small items) and the demand by individual customers is so low that some have no demand during a headway, their stops can be skipped.

For some demand processes, the proportion of stops that can be skipped should decrease with H as $\exp(-H/H_0)$, where H_0 is a constant that depends on t and $\mathbf{x}$. If the customers in a subregion are alike and their demand is well described by Poisson processes, then the parameter H_0 is the average time between successive demands at one destination; i.e., $H_0 = D'^{-1} = \delta/\lambda$. For other processes the relationship is similar. See problem 4.4 for an explanation.

As a result, the effective density of stops is only $\delta[1 - \exp\{-H/H_0\}]$. This expression must be substituted for the parameter δ in the expressions for $(\delta\alpha_2)$ and $(\delta\alpha_3)$ appearing in (4.24) (remember that δ also appears in α_2 and α_3). The optimization and design process can be carried out as described earlier. Although the resulting optimization is slightly more complicated, two extreme cases are quite simple.

First, if $H >> H_0$ then the density of stops is δ as before; the solution does not have to be changed. The opposite extreme case with $H << H_0$, arising for example if $\delta \rightarrow \infty$ but $D' \rightarrow 0$, also admits a simple expression for the stop density, even if the demand varies across customers (see problem 4.4). The expression is $\delta H/H_0 \approx \lambda H$ if items are not demanded in batches; then the number of vehicle stops per tour, $(\lambda H)A$, equals the vehicle load, λHA, as one might expect.

This limiting case arose in the study of Burns et al. (1985) involving the distribution of finished automobiles from a manufacturing plant to a very large and scattered dealer network. It is also likely to arise with flexible routing passenger transport systems as in the airport limousine model proposed by Banks et al. (1981).

4.6.2 Random Demand: Uncertain Customer Requests

If α_3 is small (items are cheap) we have seen that the minimum of (4.24) will be such that $\lambda AH = v_{max}$. There is an incentive to dispatch totally full vehicles. Let us now see what modifications are needed if the exact demand on a vehicle route is not known accurately when the vehicles are dispatched — a case with expensive items is not considered here because if time is of the essence, it is unlikely that one would operate with imperfect information. (Problems with uncertain demand have been studied in Golden and Yee, 1979; see also the review in Gendreau et al., 1996.) The system of interest operates with a headway (e.g., daily, weekly, etc.) to be determined, and advertised to customers as a service schedule that is to be met even if the volumes to be carried change with every headway. This scenario can arise for both collection and distribution problems, although for distribution problems of destination-specific items the demand will normally be known. The problem is then easy. If the size of each delivery, v_n, is both known and small compared with v_{max} it should not be difficult to partition the service region into delivery districts of nearly ideal shape with $\sum_n v_n \approx v_{max}$. Then, the distance formulae of Sec. 4.2 hold and Eqs. (4.24) can be used without modification. If some delivery lots are comparable to the vehicle's capacity, the routing problem is more difficult because one needs to balance the incentive for filling a vehicle by delivering a lot of the right size to an out-of-the-way customer with the extra distance that one would have to travel. Hall et.al. (1994) have explored improved routing schemes for this case and conclude that the basic distance formulae of Sec. 4.2 do not need much of a correction.

   In view of the above, our discussion is phrased in terms of collection, although hypothetical distribution problems with uncertain demand would be mathematically analogous. For collection problems some of the vehicles may be filled before completing their routes, which would cause some of the demands to go unfulfilled.

   The overflow customers (still needing visits) could be covered in the same headway by collection vehicles with unused cargo space or, failing that, by vehicles dispatched from the depot. Clearly, if some vehicles can be rerouted before returning to the depot, some distance can be saved. Dynamic routing introduces modeling complexities that will be discussed in Sec. 4.6.3. For now we assume that all the overflow customers are visited by a separate set of secondary vehicle routes based at the depot and planned with full information.[5] This information is gathered by the original (primary) vehicles, which are assumed to visit all the customers. Because items are "cheap" secondary vehicles should also travel full.

The decision variables are A and H , as before, but now the capacity constraint must be replaced by an overflow cost which depends on A and H . A new trade-off becomes clear. If the average demand for a tour satisfies $\lambda AH \ll v_{max}$, then the overflow cost will be negligible but most primary vehicles will travel nearly empty. On the other hand, if $\lambda AH \approx v_{max}$, a larger number of customers will overflow on average—the actual number will depend on the variability of demand as captured by its index of dispersion, γ.

Instead of a total cost per item, we work with a cost per unit time and per unit area. For given A and H , the transportation cost per unit time and unit area for primary tours is approximately independent of the overflow; it is well approximated by the product of the constant factor, λ , and the first two terms of (4.24a):

$$\frac{\alpha_1}{AH} + \frac{\delta \alpha_2}{H}.$$

Strictly speaking, this expression is an upper bound because it ignores the local delivery distance that it is saved by the stops that are skipped.

Note that, especially when the fraction of tours overflowing is small, the overflow customers will tend to be geographically distributed in widely spaced clusters of customers corresponding to overflowing tours. Because the overflow transportation cost formulas with clustered destinations are more complicated, two simple bounds will be used instead to approximate the secondary distance traveled. (Blumenfeld and Beckmann, 1984, have developed formulas for VRP's with clustered demand points). It should be intuitive without a formal derivation that smearing the clusters uniformly over **R** increases the distance traveled, while collapsing them into a single point decreases it. Upper and lower bounds for secondary distance are derived below, imagining that clusters are either spread or fused in this manner.

---

[5] An alternative approach proposed in Bertsimas and VanRyzin (1991) and also explored in Bertsimas et. al. (1991) and Hall (1992) consists in building a traveling salesman tour for the region, partitioning it into segments that do not violate the vehicle capacity constraint and then connecting the extremes of each segment to the depot by line-haul legs. Of course, for this approach to be feasible it must be possible to delay service for the kth TSP segment until the $(k-1)^{th}$ is finished. The approach is appealing because it eliminates overflows completely. On the other hand, it requires longer line-haul connections to the depot than the strategies one would use if overflows are handled with secondary tours.

An expression for f_0 , the fraction of items that must be delivered or collected as overflow, will be derived shortly. Assume for now that it is given. Then the number of secondary (overflow) tours per unit area is $\lambda H f_0/v_{max}$, and the number of stops is close to $f_0\delta$. This expression implies that the fraction of items overflowing is the same as the fraction of customers; the expression is exact if primary vehicles don't deliver (or collect) partial lots, and is also a good approximation in other cases.

With de-clustered overflowing customers, the upper bound to the secondary distance per unit area is thus:

$$\frac{2r\lambda H f_0}{v_{max}} + k\left(f_0\delta\right)^{1/2}.$$

[We are assuming here that the total number of customers is greater than the squared number of stops per vehicle: $Nf_0 \gg (v_{max}\delta/\lambda H)^2$] . With perfectly clustered groups the density of stops equals the density of incomplete primary **tours**. If we let g_0 denote the probability that a tour overflows, then this density is g_0/A ; thus a lower bound for the distance per unit area is:

$$\frac{2r\lambda H f_0}{v_{max}} + k\left(g_0/A\right)^{1/2}.$$

The secondary transportation cost per unit area and unit time is obtained by multiplying either distance bound by c_d/H , and adding to the result the cost of stopping. For the upper bound we have:

$$\begin{pmatrix} \text{overflow} \\ \text{transport} \\ \text{cost} \end{pmatrix} \approx c_d\left\{\frac{2r\lambda f_0}{v_{max}} + \frac{k\left(f_0\delta\right)^{1/2}}{H}\right\} + c_s\left\{\frac{\lambda f_0}{v_{max}} + \frac{f_0\delta}{H}\right\}$$

$$= \alpha_1\left(\frac{\lambda f_0}{v_{max}}\right) + \frac{k\left(f_0\delta\right)^{1/2}c_d}{H} + \frac{f_0\delta c_s}{H}.$$

For the lower bound, the factor $(f_0\delta)^{1/2}$ of the second term should be replaced by $(g_0/A)^{1/2}$. If the overflow is so small that only a few secondary tours are used, $Nf_0 < [v_{max}\delta/\lambda H]^2$, then k should be replaced by k' and r should be set to 0 , regardless of position.

Either on primary or secondary tours, items reach the destination at regular intervals, as required, approximately H time units apart. Thus, the stationary holding cost per unit time and unit area is:

$$\left(\begin{array}{c} holding \\ cost \end{array}\right) \approx c_h (\lambda H).$$

We are now ready to write the logistic cost function for our problem. In practical situations one would expect the difference between the upper and lower bound to be small. Therefore, we will use one of these bounds (the upper bound) below. In terms of total cost per unit time and unit area (the sum of primary and secondary transportation costs, plus the holding cost), the upper bound is:

$$\lambda z \cong \frac{(\alpha_1)}{AH} + \frac{(\delta \alpha_2)}{H} + \left(\alpha_1 \frac{\lambda}{v_{max}}\right) f_0$$

$$+ \left(k \, \delta^{1/2} c_d\right) \frac{f_0^{1/2}}{H} + \left(\delta \, c_s\right) \frac{f_0}{H} + \left(\lambda c_h\right) H, \tag{4.25}$$

where the parenthetical items are constants and the rest (A , H , and f_0) are decision variables. Note that the constant handling cost, α_0 , has been omitted from the LCF.

The fraction of items that overflow is related to A and H. As indicated by Eqs. (4.23) the mean and variance of the number of items to be carried by a primary vehicle are λAH and $\lambda A\gamma H$. The expectation of the excess of this random variable over v_{max} is the average overflow for the vehicle. Assuming that the demand is approximately normal, and letting Φ denote the standard normal cumulative distribution function (and Φ' its derivative – the probability density function), we can therefore write:

$$f_0 \cong \frac{1}{\lambda AH} \int_{v_{max}}^{\infty} (x - v_{max}) d\Phi\left(\frac{x - \lambda AH}{(\lambda A\gamma H)^{1/2}}\right)$$

$$= (\lambda AH/\gamma)^{-1/2} \psi\left[\frac{(\lambda AH - v_{max})}{(\lambda A\gamma H)^{1/2}}\right], \tag{4.26}$$

where

$$\Psi(z) = \int_{-\infty}^{z} \Phi(w)dw = \Phi'(z) + z\Phi(z),$$

which is a convex function increasing from zero (when $z \to -\infty$) to ∞ (when $z \to \infty$). Note that f_0 may depend on position and time.

Thus, Eq. (4.25) should be minimized, subject to (4.26). The procedure is simple. Conditional on AH, i.e. on the average vehicle load per district, f_0 is fixed and (4.25) only depends on H; the optimal headway can be obtained in closed form from (4.25) as an EOQ trade-off involving the 2nd, 4th, 5th and 6th terms of that expression. The resulting cost is only a function of AH, which can be minimized numerically. The procedure also works for the lower bound, and when the number of secondary tours is low. For the lower bound one should replace the fourth term of (4.25) by $kc_d(g_0/A)^{1/2}/H$, where $g_0 = \Phi(z)$.[6] Note that g_0 is fixed if AH is fixed, like f_0.

Cost estimates and guidelines for the construction of a detailed strategy can be obtained as usual, by repeating the minimization for a few combinations of (t, x). As an exercise, the reader may want to solve this minimization problem for some representative values of the input data of Problem 4.5. More ambitiously, the reader could also verify that the final strategy and the resulting cost do not change much if the overflow local distance term is replaced by the lower bound. See problem 4.5.

4.6.3 Dynamic Response to Uncertainty

In many applications, vehicle routes can be adjusted dynamically during the course of operation. For example if a collection truck of an express package carrier falls behind schedule, central dispatch can reassign some of its remaining customers to currently underutilized trucks. If a firm can do this systematically with an efficient control strategy, it should be able to operate with fewer vehicles.

To design such a system we must make a single set of planning (or configuration) decisions at the beginning of the planning period, e.g., choosing the number of trucks; and then a stream of control decisions that change dynamically as information is revealed over time. To minimize the combination of fixed and operating costs, configuration decisions must anticipate and accommodate the long-run needs of the control strategy; that is, the

---

[6] The reader may complain that the result is not a lower bound because the first two terms of (4.25) overestimate the primary local distance, by an amount $kf_0\delta^{1/2}$. However, this is also the amount that was ignored for the secondary tours when it was assumed that overflowing points within a zone were "fused" together.

system should be *planned for control*. This is difficult to do exactly but as, explained in Daganzo and Erera (1999), can be achieved approximately if we can find a family of control strategies that is: (i) parametrizable (describable in terms of just a few parameters); (ii) appealing (containing for every reasonable system configuration a near-optimal strategy for the configuration); and (iii) simple (with a predictable expected cost). Properties (i) and (iii) guarantee we can write a logistic cost function that captures approximately all fixed and recurring costs in terms of the configuration variables and control parameters. Property (ii) guarantees that good control parameters exist for every reasonable configuration. Hence, the minimum of the LCF is an "appealing" plan. Since an analytic expression exists the minimum can be searched effectively with conventional optimization methods, even if the number of variables and parameters is considerable.

The selection of a proper family is more an art than a science. The temptation is always to look for the most efficient control strategies, excelling at (ii), even if they fail the simplicity test (iii). The problem with this approach is that a search for the optimum configuration cannot then easily incorporate the effects of control. The result can be gross sub-optimization. Thus, for planning purposes we prefer to look for idealized (less efficient) control strategies that can be systematically analyzed. This allows us to explore a much larger solution space when configuring the system. The idealized strategies play the role of approximations to the more refined strategies during the optimization process, but the refined strategies can still be used when the system is operated. Several illustrations of these ideas can be found in Erera (2000). The example below is extracted from this reference.

Let us consider again the load-constrained system of Sec. 4.6.2, but assume now that H =1 day as in package collection systems. We want to configure a system where vehicles that are partially filled at the end of their runs can cover the overflow customers of other vehicles. Although very complex dynamic routing strategies can be designed to achieve this goal, we shall be satisfied with a simple one that is obviously sub-optimal but improves significantly on the static approach of Sec. 4.6.2.

We partition the service region into an inner region close to the depot (region 2) and an outer fringe (region 1). Only customers in region 1 are allocated to primary tours. We use only one planning variable: the number of primary service zones in region 1, which equals the number of vehicles, m. The radius of the inner region, r_T, is our control parameter. The idealized control strategy has two phases with several steps indicated by the numerals of Fig. 4.7.

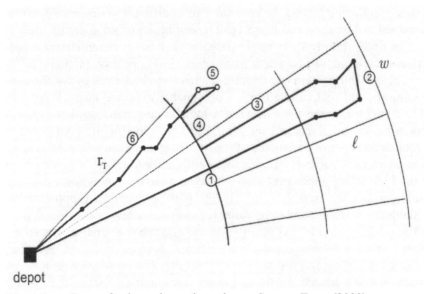

Fig. 4.7   Steps of a dynamic routing scheme. Source: Erera (2000)

In phase one vehicles travel to their service zones (step 1), serve their customers (step 2), and either return to the depot, if filled, or else stop at the boundary between regions 1 and 2 (step 3). Unfilled vehicles wait there for the start of the second phase, until all vehicles are done. Then, they are repositioned along the boundary in anticipation of serving carefully designed groups of remaining customers (step 4). The size of these groups is chosen to be consistent with each vehicle's available capacity. Vehicles first serve the part of their group in region 1(step 5), then the part in region 2 (step 6). Region 2 customers are arranged in wedges that can be served efficiently as vehicles return to the depot. Finally, if any customers remain unserved, they are served with a set of secondary tours (step 7). Note that virtually no customers require such secondary tours when systems are configured optimally.

This strategy generalizes the static procedure of Sec. 4.6.2, since the effects of the latter can be essentially achieved by setting $r_T = 0$. Although the new strategy is sub-optimal, it has clear efficiencies over the static procedure; thus, it is "appealing" in the sense of (ii). The strategy also has properties (i) and (iii), since t is parametrized by the inner radius r_T and is simple. An analytic approximation for the LCF is given in Erera (2000). The approximations in this reference were designed to be most accurate for intermediate values of r_T, where the optimum was expected to be. The formulae are not given here because they would take too long to explain, but the qualitative results are interesting.

Figure 4.8 shows how the approximate total distance per day varies as a function of r_T for a test problem, after the number of vehicles, m, was optimized. The figure also includes a dotted line from a simulation that used the recommended values of m and r_T, and a more sophisticated control algorithm. This curve gives the actual distance that could be expected in an implementation. Reassuringly, the value of r_T recommended by the optimization (the minimum of the solid line) yields a near-minimum actual distance. Note from the figure that this distance is considerably smaller than that achieved with the static strategy ($r_T = 0$). Erera (2000) shows with a battery of 20 problems that the reduction in the required number of vehicles is even greater. The portion of the vehicle fleet required by uncertainty (the "fleet penalty" in Erera's lingo) was reduced by 50% or more in 19 out of 20 cases and by more than 70% in half of the cases. The median reduction in the "distance penalty" due to uncertainty, on the other hand was only about 30%.

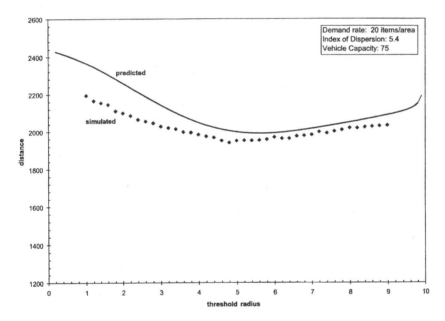

Fig. 4.8   Distance traveled for a test problem. Source: Erera (2000).

4.7 Different Customers: Asymmetric Strategies

This section explores the advantages of offering different service levels to customers with different consumption rates and/or different holding costs. Because these differences are likely to be most notable for collection problems, our discussion will be phrased in these terms – factories and manufacturing plants typically consume a wide selection of parts and raw materials even if their product line is homogeneous. Before explaining how asymmetric collection strategies can be designed, their desirability is introduced with a very simple example with two customer types, adapted from Daganzo (1985a).

4.7.1 An Illustration

Consider a problem with stationary conditions (i.e. λ and δ independent of time) obeying formulation (4.24) for which it is desirable to fill the vehicles. More specifically, we assume that: (i) the third (pipeline inventory) term of Eq.(4.24a) can be neglected because items are "cheap", and (ii) that only constraint (4.24b) plays a role because storage room at the origins is plentiful and the customer density is so large that the ideal number of vehicle stops is sure to exceed 1. We also assume that the stop cost c_s can be neglected.

Let us now examine how the optimal system cost depends on λ and δ. Because Eq. (4.24a) decreases with A for any H, its minimum is reached for as large a district area A as possible. Therefore, as expected, the vehicle capacity constraint (4.24b) must hold strictly: $A = v_{max}/(\lambda H)$. On making this substitution and minimizing the resulting EOQ expression with respect to H, a simple formula for the cost per item, z^*, is obtained. If α_4 is replaced by its expression in terms of δ and λ (i.e., $\alpha_4 = c_h\delta/\lambda$), and the result is expressed in cost units per unit time and unit area, the formula becomes:

$$\lambda z^* = \beta_1 \lambda + \left(\beta_2 \lambda^{1/2}\right)\delta^{1/4}, \tag{4.27a}$$

where $\beta_1 = \alpha_0 + \alpha_1/v_{max}$ and $\beta_2 = 4c_hc_dk$. Notice that Eq. (4.27a) increases at a decreasing rate with λ, δ and β_2; this concavity encourages discrimination, as we shall now see.

Suppose that there are two customer types, $n = 1, 2$, with demand characteristics (λ_n, δ_n) and with different c_h, so that β_2 is different for the two customer types: $\beta_2^{(1)}$ and $\beta_2^{(2)}$. (In this section we use n to index customer classes, instead of customers.) Note then that $\lambda = \lambda_1 + \lambda_2$ and $\delta = \delta_1 + \delta_2$.

If the two customer classes are treated completely separately, as if the other did not exist, the combined cost per unit time and unit area, instead of being given by (4.27a), would be:

$$\lambda z^* = \sum_{n=1}^{2} \left(\lambda_n \beta_1 + \left(\beta_2^{(n)} \lambda_n \right)^{1/2} \delta_n^{1/4} \right)$$

$$= \lambda \beta_1 + \sum_{n=1}^{2} \left(\beta_2^{(n)} \lambda_n \right)^{1/2} \delta_n^{1/4}.$$

(4.27b)

This second strategy is best if:

$$\sum_{n=1}^{2} \left(\beta_2^{(n)} \lambda_n \right)^{1/2} \delta_n^{1/4} < \left(\sum_{n=1}^{2} \beta_2^{(n)} \lambda_n \right)^{1/2} \left(\sum_{n=1}^{2} \delta_n \right)^{1/4} ;$$

(4.28)

otherwise the symmetric strategy is best.

If the two customer types are similar, Eq. (4.28) does not hold. Therefore, a symmetric strategy is best: items should be shipped together because with the higher demand density resulting from amalgamation vehicle tours can cover smaller zones and save operating costs. This is not always the case, however.

Inequality (4.28) will hold if one set of suppliers is highly concentrated ($\delta_1 \approx 0$) while producing many items that are expensive to store ($\lambda_1 \beta_2^{(1)}$ large) , and the other set has opposite characteristics (δ_2 is large but $\beta_2^{(2)} \approx 0$) . Separate service for the two sets is then reasonable because the distribution strategies for both sets should be different. For the second set one would like to save operating costs at the expense of holding cost (one would use a large H in order to reduce the area served by each vehicle) and for the first set one would do the opposite. In both cases the local operating costs plus the holding cost (left side of (4.28)) would be close to zero. However, if both items types are combined together, neither of the factors on the right side of (4.28) is close to zero – service has to be moderately frequent because some of the items are expensive to store, and tours must cover moderate size areas because all destinations have to be visited. Clearly, the requirements of the two sets of customers interfere with each other, increasing cost dramatically.

This phenomenon explains why, in real life, separate logistic systems are used to carry widely different items, even if from a transportation standpoint alone it would seem wise to combine them. It should not be surprising, thus, to find several transportation modes (taxis, limousines, buses, etc.) at the disposal of passengers exiting an airport. For freight transportation, the differences in the requirements of various customers are less likely to merit discriminating service; but the possibility should be considered.

The rest of this section describes methods for design of discriminating collection/distribution systems.

4.7.2 Discriminating Strategies

For general problems, the example just described suggests that cost may be reduced if the set of all customers is divided into classes with different characteristics, served with separate collection systems.

For a given set of classes, total cost can be easily estimated as shown in Sec. 4.6; that section described the cost and structure of near-optimal symmetric strategies, as would be used within each of our subsystems. The tricky part is defining the customer subsets that will minimize total cost. Daganzo (1985) presents a simple dynamic programming procedure to achieve this goal without detailed customer information – the method only uses the frequency (probability) distribution of customer characteristics – and shows in the process that the optimal solution would rarely exhibit more than 2 or 3 classes. When it is found that cost is minimized with only one class, discriminatory service is not cost-effective.

Although we have ignored in this section the pipeline inventory cost, and have also assumed that the same transportation mode is used for all the subsystems, this is not a prerequisite for discriminatory service to be attractive. It is impossible to discuss here all the possible cases that can arise in detail, but a general statement can be made: if customers are very different, then we should check whether dividing them into a few classes with (highly) different characteristics – and serving them separately – can reduce cost; this is unlikely to result in much gain when customers are not very different, though. Problem 4.6 is a case in point; it is solved by examining systematically all possible allocations of customers to (two) classes. Related analyses are conducted in Klincewicz et. al. (1990) and Hall and Racer (1993).

With the approach just described, each customer class n is designed separately and is characterized by design parameters A_n and H_n. By restricting these design parameters somewhat, Hall (1985) has developed a strategy that allows customers from all classes to share the transportation fleet while being visited at different frequencies. He requires A to be the same for all customers and each H_n to be an integer multiple of the time between dispatches H; that is, $H_n = m_n H$, for an integer m_n. He assumes that vehicles are dispatched at times t = 0, H, 2H, etc..., visiting each time

$(1/m_n)^{th}$ of the customers in every class n. This allows the effective stop density, $\Sigma_n\{\delta_n/m_n\}$, to be greater than for any class alone while ensuring that individual customers are only visited every m_n dispatches; it decreases the local transportation cost.

With the help of f^0 , a variable denoting the fraction of customers served in each period, Hall's strategy can be defined without resorting to classes. Accordingly, the symbol "n" now reverts to its original meaning, indexing individual customers. We seek the optimal m_n for individual customers, as well as the optimal H and f^0 . As done at the outset, let us assume that the conditions are such that vehicles will be dispatched full.

Then, the line-haul motion cost per item is α_1/v_{max} , and does not depend on the allocation scheme for customers. The local motion cost per unit time and unit area is:

$$c_d k \frac{\left(f^0\delta\right)^{1/2}}{H} + c_s f^0 \frac{\delta}{H}. \tag{4.29a}$$

(This somewhat conservative estimate assumes that stops are randomly and uniformly distributed within subregions of **R** larger than a collection district; it may be on the high side if customers of a similar kind cluster together.) The holding cost per unit time in a subregion of unit area, **P** , is:

$$\sum_{n\in \mathbf{P}} c_h^{(n)}\left(m_n H\right)D'_n. \tag{4.29b}$$

The system can be designed with a simple decomposition method. Conditional on f^0 and H , the transportation cost (4.29a) is fixed; thus, cost is minimized by the m_n's that minimize the holding cost (4.29b). These m_n's, to be consistent with f^0 , must satisfy:

$$\sum_{n\in \mathbf{P}} 1/m_n = f^0\delta. \tag{4.29c}$$

Once the m_n have been found, the conditional total cost is obtained. Testing various values of f^0 and H , we can identify a near-optimal solution. Alternatively, if one replaces the constraint $[m_n = 1, 2, 3, ...]$ by $[m_n \geq 1]$, a simple approximation for the minimal holding cost for a given f^0 and H can be obtained (see Problem 4.7). The optimal strategy is then defined by the minimum over f^0 and H of the sum of this approximation and the local motion cost expression.

4.8   Other Extensions

One of the reasons for the very extensive literature on algorithms to vehicle routing problems is that in actual applications almost every problem has some peculiarity that renders it unique. We have already seen that there can be a variety of cases depending on:

(i)   the relative size of the number of tours and the maximum number of stops per tour (Section 4.2),

(ii)   the relative cost of rent, inventory, and operating costs. (Sections 4.3 and 4.4),

(iii)   limitations to route length and storage space (Section 4.4),

(iv)   dissimilarity in the values of items and the demand rates at different destinations. (Section 4.7),

(v)   amount of uncertainty as to the customer lot sizes. (Section 4.6).

In addition (and this is not an exhaustive list) one might find situations in which time enters the problem because customers request service during certain "time windows", or there is a limit to the amount of time an item can spend in transit (perishable items). There also are situations where vehicles do *both* distribution and collection (routing with backhauls), and situations where vehicle loading considerations make it advantageous to visit customers in an order which does not minimize the total distance traveled.

It is clearly impossible within the scope of a short book to study in detail even a partial list of possible cases (and combinations of cases). But it is possible to give a broad recipe to deal with routing peculiarities. A solution to problem 4.8, an elevator system design, can be obtained with such an approach.

4.8.1   Routing Peculiarities

At the core of our proposed two-step method for solving general distribution problems there should be a simple and efficient routing algorithm, whose performance can be quantified by means of simple formulas using average density as an input, instead of detailed customer locations. It is then a simple matter to add holding and pipeline inventory costs to the motion cost to define a logistic cost function. If routing/scheduling strategies can be defined in terms of a few decisions variables that are constrained only locally in the time-space domain, then the minimum of the (constrained) logistic cost function will approximate the cost generated by items in different portions of the time-space domain. The CA approach can be used.

Some routing cost models that allow this to be accomplished already exist. They are now briefly reviewed. Simple transportation cost formulas have been proposed for time-window problems (Daganzo, 1987a,b). The results show how cost increases with the narrowness of the windows, and with the proportion of customers with tight requirements. The proposed routing strategy uses a different set of delivery districts for the customers in each time window, and staggers the zones in such a way so as to leave most vehicles in favorable locations at the beginning of each new window period.

Perishable items such as newspapers (Han, 1984, and Han and Daganzo, 1986), lead to VRP structures which are similar to those arising from the vehicle route length limitations discussed in Sec. 4.4.1. The main difference is that service districts that are far away form the depot should be (i) more elongated than usual and (ii) covered in a one-way pass that begins at the end of the district that is close to the depot and terminates at the far end. Although this modification increases the line-haul distance traveled, it also allows distribution to begin sooner and the districts to include more stops.

Models with both pick-ups and deliveries have been constructed for public transportation systems (Daganzo, Hendrickson and Wilson, 1977, Hendrickson, 1978) serving one focal point and a surrounding area. The strategies examined in these early works, however, are not as general as possible; they only consider two extreme cases for a partition of the surrounding area into service zones. More recently, Daganzo and Hall (1990) present an improved cost model for routing with backhauls, emphasizing cases where the total flow in one direction (e.g. outbound from the depot) is a few times larger than in the other direction. The basic idea is briefly summarized below for the case where the dominant flow is outbound; the reverse situation is similar. One simply constructs distribution tours as if there were no pickups, allocates each pickup to the nearest return leg of a distribution trip (or "spoke"), and finally modifies the vehicle tours in recognition of the newly assigned stops. Because the density of spokes increases rapidly toward the depot, significant vehicle deviations are only required for pickups near the outer fringe of the region. Pickup miles on the fringe can be reduced by ending the outermost delivery tours at the far end of their districts and by other modifications that are geared to optimize the spatial distribution of spokes. In fact, it is shown in Daganzo and Hall (1990) that under some conditions it is almost as if the secondary stops added only a stop cost and no distance cost. Hall (1993) has applied the concept of spokes to the VRP problem for deliveries only, in which customers demand large and small items.

Another complication that deserves attention involves the interaction of vehicle loading and routing. When items have awkward shapes and are large, so that only a few fit in a vehicle, v_{max} may not be fixed; it may depend on the specific customers that are visited or even the order in which they are visited. The latter phenomenon may arise if weight distribution restrictions, for example, dictate that some items (and thus some stops) must be handled before others. This topic is very complex and hard to handle generally; see Hall (1989) and Ball et al. (1995a) for example.

4.8.2 Interactions with Production

Another area where further results may be desirable involves the interaction of physical distribution with production schedules. This interaction sometimes offers an opportunity for further cost reductions.

This subject was broached in Sec. 4.3.3, where it was suggested that production of (destination-specific) items should be rotated among geographical customer regions every headway H. Dispatching the vehicles to a region immediately after its production run was completed greatly reduced the holding costs at the origin. It was assumed that production would be coordinated with transportation in this manner without much of a penalty. More likely, though, there may be a set-up cost associated with each switch in production item types. In this case production costs may be reduced by switching less frequently and holding higher inventories at the origin. An integrated solution can then be obtained by including in the logistic cost function the production set-up costs, e.g., as explained below.

If no attempt is made to coordinate the production schedule with the physical distribution schedule, then the inventory at the origin of items of a certain type can be decomposed as shown in Figure 4.9 into a (shaded) component which depends on the time between setups for that item type,[7] H_s , and a (dotted) component which depends on the transportation headway, H; see Blumenfeld et al. (1985a):

$$\left(\begin{array}{c} average\ inventory\ cost \\ per\ item\ at\ origin \end{array} \right) \approx \frac{c_i}{2} H_s + \frac{c_i}{2} H.$$

---

[7] We are assuming that the number of item types is large and, therefore, the steps of the production curve are nearly vertical. Similar conclusions can be reached for few item types.

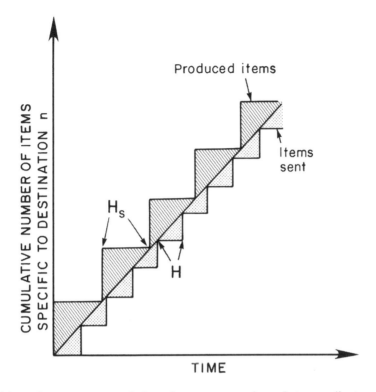

Fig. 4.9   Inventory accumulation when no attempt is made to coordinate production and distribution

The maximum accumulation also decomposes in a similar manner:

$$Maximum\ accumulation \approx H_s D' + HD.$$

Because production costs depend on H_s and not on H , the sum of the production and logistics costs is made up of two components: (i) a production component with *only* production decision variables (including H_s), and (ii) a logistic component with *only* logistics variables (including A and H). Logistics and production decisions, thus, can be made independently of each other.

By selecting H to be an integer submultiple of H_s , or vice versa, it is possible to reduce the inventory time at the origin by an amount equal to the smallest of H and H_s (Figure 4.10 depicts the case with $H_s = 3H$), and the maximum accumulation becomes the difference between the maximum and the minimum of $H_s D'$ and HD' .

If this kind of coordination is feasible, the sum of the production and logistics costs no longer decomposes, and a coordinated production and distribution scheme should be considered. Blumenfeld et. al. (1985a) and (1986) have examined the case where each district is constrained to contain only one destination and all shipments are direct ($n_s = 1$). They illustrated situations where coordination of production and distribution is most conducive to cost savings, and provided a bound on the maximum possible benefit. Further research may be worthwhile to relax the $n_s = 1$ assumption and to allow more destinations than item types.

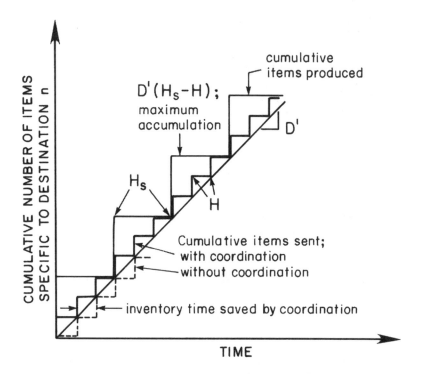

Fig. 4.10   Inventory accumulation with coordinated schedules

Throughout the chapter it was assumed that the total production rate, and not just the schedule by item type, could be adapted to the changing demand without penalty. In practice, though, this is rarely so, even if the items produced are generic. (It is more costly to change the *quantity* of items produced than the *kind* of items produced because to adjust the production rate one needs to hire extra labor, pay overtime or fire labor as needed – and the penalty for these actions is large; Newell, 1990, has examined the production rate adjustment process.) We conclude this chapter by showing that this seemingly strong assumption can often be relaxed.

Figure 4.11 shows how a production curve may be adapted to a gradually decreasing demand; the objective is tracking the smooth envelope to the crests of the shipment curve (which varies like the demand curve) as closely as possible, without many production rate changes. We had seen in Sec. 2.5 that for a similar model, portrayed in Fig. 2.10, lot size decisions were independent of production decisions; fortunately, this is also true now. In Fig. 4.11, the inventory at the origin decomposes in two components: (i) a (shaded) component, which is due to the discreteness in the production rate changes and is independent of the shipping schedule,[8] and (ii) a dotted component which is the same as if the production schedule was adjusted continuously as assumed in this chapter. Thus, costs can be divided into two components affected respectively only by production, or only by logistics decision variables.

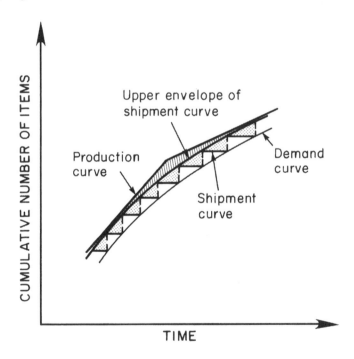

Fig. 4.11   Production for a gradually decreasing demand

---

[8] It is the same as if the production curve was driven by the demand curve itself.

Suggested Exercises

4.1   The maximum number of stops made by a vehicle delivering lots of size v (v < v_{max}) to identical customers is $[v_{max}/v]^-$ if a delivery lot cannot be split among vehicles. If v_{max}/v is an integer, an alternative way of expressing Eq. (4.5a) is:

$$Total\ distance \cong \left\{ 2E(r)v/v_{max} + kE\left(\delta^{-1/2}\right) \right\}\ N.$$

If v_{max}/v is not an integer this expression is a lower bound. The bound is very tight if delivery lots can be split among vehicles in order to fill them. Hall (1993) explores in detail the shipment splitting issue.

(i)   Show that this expression also applies if $v > v_{max}$ and all the trucks are dispatched full (splitting delivery lots as needed), provided that a condition (equivalent to $N \gg C^2$) holds. Write the condition. Show as well that the expression is a (tight) lower bound if delivery lots smaller than v_{max} are not split.

(ii)   Derive as well the generalization of Eq. (4.6b), when the condition does not hold. Show that the distance expression is close to the above.

4.2   Section 4.4.1 describes how to design a one-to-many distribution system when the maximum number of stops per tour cannot exceed $C_{max}(\mathbf{x})$, a quantity that depends on location. Develop an expression for $C_{max}(\mathbf{x})$ when, aside from vehicle capacity, the only restriction to number of stops is the maximum time allowed for a vehicle tour, t_{max}. (Assume that the vehicle's average moving speed and time per stop are known.)

4.3   If Eq. (4.21) is a good representation of the optimal cost for all (t, $\mathbf{x}$), explain the logic behind the following expression for the average cost over $\mathbf{R}$, for the time horizon (we assume that $c_s \approx 0$):

$$average\ cost\ per\ item = 2(r)c_d/v_{max}$$
$$+ 2(kc_hc_d)^{1/2} E\left(D'^{-1}\right)E\left(\delta^{-1/4}\right)$$

If D' and δ do not change much across items – i.e. their standard deviation to mean ratios (denoted here by ε and ε') are small compared with one – show that:

$$average\ cost\ per\ item \approx 2E(r)c_d/v_{max} +$$

$$[\ 2(kc_hc_d)^{1/2}\ t_{max}/\{D(t_{max})\ [N/|\mathbf{R}|\ ]^{1/4}\}\ ]\ [\ 1 + \varepsilon^2 + \varepsilon'^2\ ]$$

This expression reduces to the above if $\varepsilon = \varepsilon' = 0$, and like the above it increases with the ε's .

[**Hint:** Use a well-known approximation to the expectation of a non-linear function of a random variable, based on a two-term Taylor series expansion of the function.]

4.4   Assume that the demand for items by every customer in $\mathbf{R}$ follows a compound Poisson process with an inex of dispersion γ; recall that the average number of events occurring per unit time is $D'_c = D'/\gamma$. Then explain why the proportion of all the customers in a subregion of $\mathbf{R}$ with no demand in time H is $\exp[-D'_cH]$. (Note that the constant H_0 defined in the text is the inverse of D'_c ; as the average time between successive events, it is: $H_0 = \gamma/D' = \gamma\delta/\lambda$.)

4.5   Assume that service is to be provided to 400 customers uniformly scattered in a (20 mi. x 20 mi.) square region, with the depot in a corner. The following numerical constants describe the situation: $\lambda/v_{max} = 0.1$ (truckloads/day-mile^2); $\alpha_1/r = 2$ ($/ vehicle-mile) ; $\alpha_2 = 21$ ($/customer visit) ; $kc_d = 1$ ($/veh-mile) ; $c_s = 20$ ($/vehicle stop); $c_h\ v_{max} = 30$ ($ / truckload-day) ; $\gamma = 1$ (truckload of items). Solve the minimization problem defined by Eqs. (4.25 and 4.26), estimate the upper and lower bounds to total cost, and define a distribution strategy for the primary and secondary tours.

4.6   Items are to be carried from a depot to many scattered destinations. Two modes of transportation are available: mode A is fast $s_A = \infty$ and expensive $c_d^A \gg c_d^B$. Mode B is slower: $s_B < \infty$. These features aside, the rest of the features are the same for both modes. They share the cost per stop, vehicle capacity and loading/handling cost.

Storage space at the origin and the destinations is plentiful, so rent costs are neglected, but inventory cost changes drastically across destinations. This is described by a cumulative density function which gives the density, $\delta(c_i)$, for all the destinations with carrying cost below c_i. (For any c_i these are assumed to be uniformly and randomly scattered about the service region, **R**.) If the demand is stationary and the same at all destinations, describe qualitatively a procedure for allocating the destinations to modes and the optimal service characteristics (A and H) of each mode. Assume that all the customers allocated to the mode are served in every headway, and that the pipeline inventory cost *cannot* be neglected.

[**Hint:** Prove that if destination n_A goes on mode A and n_B goes on node B, then the inventory cost of n_A must be greater than the inventory cost of n_B . Then find the minimum cost for both transportation modes as a function of a critical $\tilde{c}_i$; iff $c_i < \tilde{c}_i$, then mode B is used.]

4.7   Derive a closed form solution for the minimum of Eq. (4.29b) subject to (4.29c). Assume without loss of generality that the n are arranged in order of increasing G_n [where $G_n = (D'_n)(c_h^{(n)})$] , and that the units of measurement are such that $\delta = 1000$, and $c_s = 1$.

Then, if the fraction of customers with G_n below x , $P(x)$, is:

$$
\begin{aligned}
P(x) = \quad & 0 && \text{if } \ x \le 1 \\
= \quad & a(x\text{-}1) && \text{if } \ 1 < x < [\,1 + a^{-1}\,] \\
= \quad & 1 && \text{if } \ x \ge [\,1 + a^{-1}\,]
\end{aligned}
$$

solve for the minimum f^0 and H . Do this for various values of "a" and describe the result.

4.8   An 80 story office building is served by 20 elevators. Whether full or empty, elevators travel at a speed of one floor per 0.5 seconds. Each time they stop at a floor, their travel time is increased by Δt seconds:

$$\Delta t = 10 + n,$$

where n is the number of passengers exiting or entering the elevator at that particular floor.

In the morning rush hour all the traffic originates at the lobby, at a constant rate of two passengers per minute per floor.

1. Arrange the elevators in banks so that the average passenger waiting plus riding time is minimized. (A bank is a set of "m" elevators serving the lobby and "b" contiguous floors.) Ignore pairing and assume that the arrival rate is maintained for a long time. Assume as well that an elevator can hold as many people as needed.

2. For part 1, it was reasonable to expect every elevator to stop at every floor. During the off-peak, however, (with demand rate λ $\ll 2$ pax./min.) elevators may skip floors. Discuss how the optimal banking strategy is affected by λ . As an aid for thinking, you may use the following approximate expression for the total service time, $T(x)$, for a passenger going to floor x (a floor that is included in a bank with b floors and is served by m elevators):

$$T(x) = \left[\left(1+\frac{1}{m}\right)\frac{R}{2}\right] - \left[\frac{(b-1)S}{2}\right],$$

where S is the elevator speed in floors/sec., and R is the elevator round trip time for that bank,

$$R \approx \left(\frac{k-15m}{\lambda}\right) + \left[\left(\frac{k+15m}{\lambda}\right)^2 - \left(\frac{6000mb}{\lambda}\right)\right]^{1/2} \quad secs.,$$

with k = 5(1 + b) + [x + (b-1)/2]/S .

This expression assumes that $\Delta t = 10$, regardless of n. It can be included in a spreadsheet for sensitivity analysis. Try $\lambda = 0.1$, 0.3 , and 1.0 pax/min. Note the influence that elevator speed has on the homogeneity of your bank configuration.

3. As a prelude to the material in Chapter 5, discuss how the use of skylobbies could improve the level of service and reduce the amount of floor space consumed by elevator shafts.

[**Hint** for part 1: The number of elevators per floor should be about the same for each bank. Note that this is an engineering design problem; if you cannot identify the optimum analytically, you must still propose the best design possible. Use of a computer spreadsheet is recommended.]

Glossary of Symbols

α_0:	Handling and fixed pipeline inventory cost (\$/item),
α_1:	Fixed cost per vehicle dispatch (\$/dispatch),
α_2:	Transportation cost added by a customer detour (\$),
α_3:	Pipeline inventory cost added by a customer detour (\$/item),
α_4:	Stationary holding cost for one item during the time between demands (\$/item²),
A, A(x),A(t,x):	Area of **A**,
a:	Number of aisles with a request,
A_n:	Area of delivery district used for type-n customers,
A:	Subregion of **R** ; delivery district,
ß:	Characteristic constant used in Sec 4.4.2,
$ß_1$, $ß_2$, $ß_2^{(n)}$:	Constants used in Sec 4.7.1,
c_d:	Cost per vehicle-"mile",
c_f:	Terminal handling cost constant,
c_h:	Holding cost per item-day,
$c_h^{(n)}$:	Holding cost per item-day for type-n items,
c_i:	Inventory cost per item-day,
c_r:	Rent cost per item-day,
c_s:	Fixed transportation cost of a vehicle stop,
c'_s:	Added transportation cost of carrying an extra item,
c_r:	Rent cost per item-day,
c_t:	Vehicle operating cost,
C, C_{max}:	Maximum number of stops made by a vehicle,
$C_{max}(x)$:	Maximum number of stops in the neighborhood of **x**,
C_p:	Maximum number of stops in the pth subregion,
δ, $\delta(x)$, $\delta(t,x)$:	Spatial customer density (customers/area),
δ_n:	Spatial density of type-n customers (customers/area),
d:	Tour length,
$\overline{D}\,'$:	Time averaged demand rate for a customer,
D', D'(t):	Demand rate per customer (at time t),
D_{mp}:	Demand in subregion "p" during time interval "m" ,
$D_n(t)$, D(t):	Cumulative demand of customer "n" by time "t",
ε:	A mathematical ratio in the full vehicle condition proof,
E( ):	Expectation of a random variable,
Φ( ):	Standard normal cumulative distribution function,
f(**x**):	Probability density (for customer location) at **x**,
f_0:	Fraction of items collected as overflow,
γ, $\gamma(t,x)$:	[Variance/Mean] of items demanded in a time interval (index of dispersion),
g_0:	Probability of overflow,

H, H(t), H(t,$\mathbf{x}$): Headway,

H_0:   Time constant (in Sec. 4.6.1 only),

H_ℓ: ℓth headway,

H_n: Headway for the nth customer class,

H_s: Time between production setups,

k: Dimensionless factor for the VRP local distance; (more vehicle tours than stops per tour),

k':   Dimensionless factor for the VRP local distance; (fewer vehicle tours than stops per tour),

λ , $\lambda(t,\mathbf{x})$:   Demand density rate (items/time-area),

λ_n: Demand density rate for type-n customers (items/time-area),

$\ell = (1,...,L)$:   Indexes for dispatching times and headways,

L_p: Number of dispatches for the pth subregion,

$m = (1, ...,M)$:   Indexes for time intervals,

$n = (1,...,N)$:   Indexes for customers, destinations, and customer classes,

n_s, $n_s(t,\mathbf{x})$:   Number of stops per tour , at $(t,\mathbf{x})$,

n_s^*:   Optimal number of stops per tour,

$n_{s,i}$:   Number of stops per tour, for tours near x_i,

n_s^ℓ (and $n_s^\ell(\mathbf{x})$): Number of stops per tour (near $\mathbf{x}$) for the ℓth dispatch,

N_p: Number of destinations in the pth subregion,

$p = (1,...,P)$:   Indexes for the subregions of $\mathbf{R}$,

$\mathbf{P}_p$:   pth subregion in a partition of $\mathbf{R}$,

r: Average distance from the points in a delivery region to the depot,

$r(\mathbf{x})$, (or r_i):   Distance from the depot to point $\mathbf{x}$, (or x_i),

r_p: Average distance from the pth subregion to the depot,

$\mathbf{R}$:   Service region,

$|\mathbf{R}|$:   Surface area of $\mathbf{R}$,

s :   Vehicle speed,

$\tau(\mathbf{x})$:   Time to cover a unit area around $\mathbf{x}$. (Sec. 4.5.1 only),

τ_m: The mth time interval in a partition of the study period,

t :   Time,

t_ℓ: Time of the ℓth dispatch,

t_m: Average time spent in a vehicle (per item),

t_{max}:   End of the study period,

t_s: Time duration of a vehicle stop,

$T(\mathbf{x})$:   Express time to point $\mathbf{x}$. (Sec. 4.5.1 only),

v (and v_ℓ):   Delivery lot size to a customer (for the ℓth shipment),

v_{max}:   Vehicle capacity (items),

v^o, $v^o(\mathbf{x})$:   Maximum allowable accumulation at a destination (items),

$\mathbf{x} = (x_1,x_2)$:   Spatial coordinates of a point,

x_o: Inner point of $\mathbf{A}$,

z: Cost per item,

z*:   Optimal cost per item,

z^h and z^ℓ: Upper and lower bounds to z*. (Sec. 4.4.2),

z^{mp}:   Cost per item in subregion "p" for time interval "m",

z_m: Motion cost per item,

z_p:	Pipeline inventory cost per item,
z_s:	Stationary inventory cost per item,
$[\ ]^+$:	Closest integer from above to the argument in brackets,
$[\ ]^-$:	Integer part of the argument in brackets.

5 One-To-Many Distribution with Transshipments

Readings for Chapter 5

Designing a one-to-many logistics system with transshipments is a complex task, as one must decide how many terminals will be operated, their location, the routes and schedules of the various vehicle types operated, and the allocation of customers to specific terminals and routes. Daganzo and Newell (1986) shows how the design problem can be reduced to a simpler terminal sizing and location problem, as explained in Sec. 5.2. This reference is also at the core of the design discussion in Secs. 5.3 and 5.5. The discretization approach presented in Sec. 5.6 is taken from Ouyang and Daganzo (2004).

5.1 Initial Remarks

This chapter introduces the possibility of transshipments for one-to-many distribution (or collection problems). A transshipment is the act of taking an item out of a vehicle and loading it onto another. Typically, transshipments take place at fixed facilities, which we call terminals. For modeling purposes, these can be viewed as a set of berthing gates connected by an internal sorting, storage and transfer system. The berthing gates accommodate the vehicles while they are being loaded and unloaded; the sorting-storage-transfer system moves the items from one vehicle to another. Although many different technologies exist depending on the freight that is being moved, conceptually this makes little difference. (The internal transfer system, for example, can use: carts on rails, forklift trucks, conveyor belts, idler rollers or gravity chutes.) The emphasis at efficient terminals is on moving the freight quickly with little allowances made for long term storage. But if there is a need to accommodate seasonal fluctuations in demand, or to hold inventories closer to the points of demand when response time is critical and demand cannot be anticipated, terminals can also provide a warehousing function.

The chapter is organized as follows. Section 5.2 introduces qualitative properties of near-optimal systems, which allow the problem to be treated analytically; the remainder of the chapter uses these results. Section 5.3

shows in detail how systems where items are transhipped no more than once can be designed, using an uncomplicated scenario as an illustration. Section 5.4 then describes modifications to the procedure able to capture the following complicating features: schedule synchronization, variable and uncertain demand, asymmetric strategies, as well as constraints on locations, routes and schedules. Because the overall approach remains unchanged, Section 5.4 is rather concise and focuses on the specific modifications; problems at the end of the chapter can be used for further study. Finally, with the one-transshipment results as a building block, Section 5.5 solves the multiple transshipment problem. Section 5.6, then shows how to computerize the design guidelines used in Sections 5.3, 5.4 and 5.5.

5.2   Distribution with Transshipments

After reviewing the reasons for transshipments in one-to-many logistics systems, this section will show that finding the ideal spatial arrangement of terminals is the critical step in designing a system. The rest is easy because, for a given arrangement, there is a well defined set of item paths, vehicle routes, and schedules that (nearly) minimize total cost.

5.2.1   The Role of Terminals in One-to-Many Distribution

Items are often transshipped when there is an incentive to change transportation modes or vehicle types. While geographical barriers such as coastlines invariably require a modal change (e.g. at seaports), purely economical considerations may also encourage changes in vehicle type.

We saw in Chapter 4 that vehicles should be filled to capacity for the distribution of "cheap" freight; i.e., where pipeline inventory cost is negligible compared to the other logistic cost components. Because the optimal cost was a decreasing cost of v_{max}, we argued that (if there is a choice) one should use the largest vehicles that the local roads and the destination loading/unloading facilities can accommodate. If vehicle size is limited in the immediate vicinity of the customers, transshipments at terminals in the general neighborhood of the customers may be attractive, as this could allow larger vehicles to feed the terminals.

Consider Fig. 5.1a. This figure depicts one origin (the depot) and four customers that receive direct service once a day. Each *daily* trip is represented by one arrow joining the beginning and end of the trip. Let us assume that the pattern of the figure is optimal for the situation at hand, and that the trips are made by delivery vans, due to the small access roads leading to the customers.

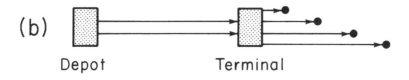

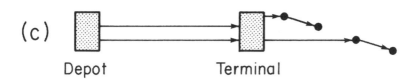

Fig. 5.1   Effect of a transshipment on vehicle-miles traveled

If a terminal is introduced, as shown in Fig. 5.1b, the transportation cost can be reduced without changing the service frequency to the customers (i.e. the waiting cost at the destination). If the main roads can accommodate trucks with twice the capacity of vans, then only two trucks need to be dispatched between the depot and the terminal every day; as a result the transportation cost can be cut by a factor of two, or close to it. Destinations can still be served daily by vans from the terminal, as shown in the figure. This arrangement clearly reduces the total vehicle-miles traveled per day (the sum of the lengths of the arrows in the figure) and does not change the holding costs at the destinations. On the other hand, the arrangement may increase holding cost at the origin – items now leave the origin in larger batches – and introduces new handling and holding costs at the terminal. Whether the distribution scheme of Fig 5.1b is advantageous will depend on the magnitude of the transportation cost savings, which grow with the distance between the terminal and the depot, and with the size difference between vehicles delivering to the terminal and the customers.

A benefit from transshipments may be derived even if, due to route length limitations, vehicles cannot travel full. (Recall from Sec. 4.4 that pipeline inventory considerations, in addition to operating restrictions such as the duration of a work shift, may restrict delivery route length; very valuable items should not be delivered on many-stop routes.) To illustrate this benefit, imagine that the system in Fig. 5.1a is optimal, and that its vehicles leave the depot only 1/2 full. In other words, we are assuming that increasing (or decreasing) the delivery lot size is not desirable because holding costs at the destination would then be too large (or too small). Furthermore, although one could presumable reduce costs by using delivery routes with two stops without changing the delivery frequency, we also assume that the loading/unloading operation is so slow that there is no time in a work shift to make more than one stop and return to the depot. Thus, without transshipments, the arrangement can be assumed to be optimal.

Clearly, the introduction of transshipments as in Fig. 5.1b allows matters to be improved, since the terminal allows the routes to be broken into shorter segments. Although deliveries still take place in half filled vehicles, the terminal is supplied by full vehicles. Further improvement is possible. Because the deliveries now start from a place closer to the destinations, there may be time to make more than one stop and reduce even more the daily distance traveled for local delivery, as illustrated in Fig. 5.1c. No change in delivery lot sizes results.

In summary, terminals allow us to decouple the line-haul transportation and local delivery operations, enabling us to use larger vehicles for line-haul than are used for delivery; they also increase the number of delivery stops that can be made without violating route length limitations. We will see in Chapter 6 that terminals can also play a "break-bulk" role for many-to-many problems.

5.2.2 Design Objectives and Possible Simplifications

The structure of a distribution system is defined by the number and location of the transshipment points, the routes and schedules of the transportation vehicles, and by the paths and schedules followed by the items. Usually, the number and location of the transshipment points cannot be changed as readily as routes and schedules. The latter are *tactical* level variables, and the former *strategic* variables. Since customers are usually not affected by routing changes, the vehicle routes and item paths can often be viewed as *operational* level variables, which can be changed even more readily than the delivery schedules.

For long term (strategic) analyses, decisions at all levels (operational, tactical and strategic) need to be made. For this type of problem we will develop optimal system configurations assuming that the terminals can be opened, closed and relocated without a penalty. This simplification is not as restrictive as it may seem because, if conditions change slowly with time, locations do not need to be changed often. If $\lambda(t,\mathbf{x})$ changes slowly with time, near-optimal terminal locations will be shown also to change slowly with time (this dependence is even more sluggish than the dependence of headways and number of stops on t , studied in Chap. 4). Because the overall cost is not overly sensitive to the specific locations (as seen in Chap. 3), one can keep a given set of terminals for a long time before some need to be opened, closed or relocated. In any case, relocation costs are likely to be greatly reduced by current trends in the logistics industry, such as the advent of "third-party logistics" firms that furnish full service terminal/warehousing facilities; see Martin (1989) for a description of this type of operation.

Unless the changes in $\lambda(t,\mathbf{x})$ arise from policy decisions (e.g. expanding the service region over time), the timing of changes to λ may be hard to predict. Without reliable information on them it might be reasonable to design the system as if the changes occurred gradually, using a smooth forecasted $\lambda(t,\mathbf{x})$ demand density, or else adapting to the current circumstances as time passes. In either case one would rarely expect the optimal distribution of terminals to change rapidly with time, and it should be possible to design a strategy for opening, closing and relocating terminals that maintains a near-optimal distribution of terminals without large relocating costs. (Campbell, 1990a, has examined a few dynamic strategies; some will be described later in the Chapter.)

For the short term one may be interested in adjustments to the tactical and operational decision variables. We may want to determine the best set of vehicle routes and frequencies for a given set of terminals; including, of course, the possibility of not using some of the terminals. These (tactical) problems will also be discussed in this chapter, although strategic analyses will be its main focus.

Obviously, the design problem is very complicated if considered with all its details. Our immediate goal, thus, will be to reduce it to a form involving little data and few decision variables, yet capturing the essence of the logistical costs. The remainder of this section is devoted to this endeavor; it describes some properties of near-optimal distribution systems with terminals that allow the formulation to be greatly simplified. The discussion is based on Daganzo and Newell (1986).

Figure 5.2 depicts a physical distribution network to carry items from one depot to multiple customers. The network includes terminals (dots on the figure), and multi-stop vehicle routes (looping arrows) that may stop at terminals and customers (x's) .

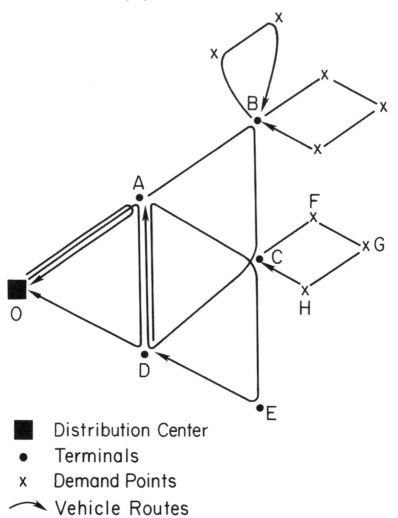

Fig. 5.2   Structure of physical distribution systems. (Source: Daganzo and Newell, 1986.)

Because we are only looking at distribution, we shall assume that a vehicle only loads items at the beginning of its route and only distributes them in succeeding stops. This is a reasonable assumption because the within-vehicle sorting complexity and stowage/restowage costs would increase substantially otherwise during a tour. Even collection/distribution systems, for which the savings of interspersing pick-ups and deliveries are obvious, tend to segregate them on individual tours.

An item that needs to be taken to destination F in Fig. 5.2 may use vehicle routes (OAO, ABCDA, and CFGHC), or (OADO, DACED, and CFGHC) to get there. In the first case, it would use path OABCF and in the second case OADACF. If redundant network structures, where some destinations can be reached by more than one path (such as those of Fig. 5.2), can be shown not to be necessary, we would like to rule them out before starting any analysis. This is done next.

Near-optimality of non-redundant networks: Here we show that, in many situations, networks providing redundant paths are not needed because total cost is concave on flow.

For the proof we focus on an operational problem, where the terminal locations, vehicle routes and schedules are fixed but one can choose the item paths and vehicle sizes. Then, the daily cost of transportation will be the sum of the transportation costs on each route. Each one of these route costs should only depend on the size of the vehicle used on the route, as per the discussion of Chap. 2. Furthermore, the relationship should be concave and increasing because of the economies of scale in vehicle size. Clearly then, on each route we should choose the smallest vehicles able to carry the load. Because the size of the vehicle must be proportional to the flow of items on the first link of its route, and these link flows are linear functions of the item path flows, transportation cost must be a concave function of the path flows. Assuming that the path of each item is chosen at the distribution center 0, independently of the time at which it becomes available for shipment and of the characteristics of the item, we see that the average time that items are waiting outside vehicles on a specific path is not affected by the path selection strategy at 0; the average time is fixed. Since travel times are also fixed, total inventory costs must be linear in the path flows. Therefore, the total distribution cost (if rent costs are ignored) must be concave in the path flows. (We recognize that rent costs are not concave. These costs, however, are typically small compared with transportation costs and should, thus, be unable to reverse the effects of concavity.)

Before discussing the implications of concavity, it is worth clarifying the two exceptions that were made in the above argument. If the selection of a path for an item is allowed to depend on the time it becomes available for shipment (e.g., passengers using public transportation systems will often choose the first of several lines to depart, if there is a choice) the stationary inventory cost depends on flow; examples can be built where total inventory cost is convex in the path flows. Even in the (rare) case where dynamic path selection is an option, it is unlikely that one would provide multiple paths to exploit such dynamics.

The second exception refers to items of different characteristics. As shown in Chapter 4, sometimes it is advantageous to send items of *widely* varying prices per unit weight on different paths (e.g., expensive items by air freight and cheap goods by land). In such cases, the pipeline inventory cost is not linear in the path flow; it depends on which items are sent on specific paths. The cost concavity argument does not hold either.

If all customers are treated alike – asymmetric strategies where this is not the case will be discussed later in the chapter – and rent costs are not dominant, then total costs are concave in the flows; in other words, there are scale economies. In this case, as we showed in Chapter 3, only one path should be used to reach each destination.

The arguments of that chapter also apply if the destination is an intermediate terminal because intermediate path flows are linear functions of path flows and concavity is preserved. Consequently, path redundancy to either intermediate or final destinations is not needed. If follows that *each terminal, or final destination point, needs to be served by only one vehicle route.* Otherwise, the stop could be bypassed by all vehicle routes carrying no flow to it for a reduction in transportation cost.

This implies that each destination point should be on only one route from only one terminal. That is, if we define the level-n influence area of a terminal as the set of points that are served from it with n or less transshipments at succeeding terminals, level 0 influence areas must form a partition of the service area. Since each terminal can only be on one vehicle route starting at another terminal, the influence areas at every level must also form a partition. Fig. 5.3 displays a possible structure where influence areas are simply connected sets (with no holes). We will reasonably assume from now on that influence areas are simply connected.

Near-optimal operations: Given the dispatching frequency from every terminal, we describe here which stops should be served from which terminals, and the structure of the vehicle routes based at each terminal.

To build such routes in a near-optimal way for a given set of stops their length should be minimized. Otherwise, a reduction in length could reduce total cost through decreases in the pipeline inventory cost, and the transportation cost. Thus, it seems logical to construct the routes with a VRP technique, as described in Sec. 4.2.

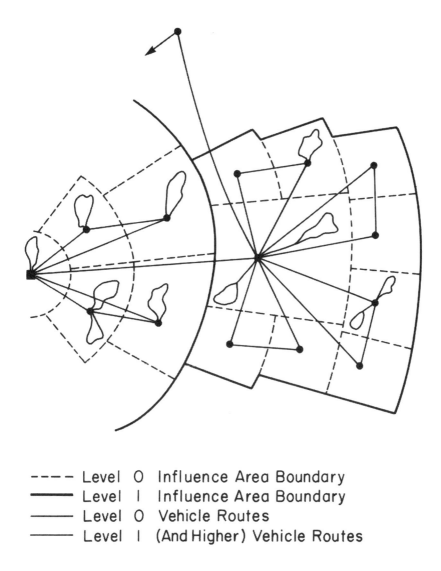

---- Level 0 Influence Area Boundary
—— Level I Influence Area Boundary
—— Level 0 Vehicle Routes
—— Level I (And Higher) Vehicle Routes

Fig. 5.3   Warehouses and influence areas. (Source: Daganzo and Newell, 1986)

We also need to decide which stops are to be served from which terminal. It will be assumed that vehicle routes do not stop at both terminal and final destinations. This is reasonable (and common practice) because otherwise sorting and scheduling work would increase substantially in size and complexity. For systems with more than one level of terminals it will be assumed that vehicle routes only stop at one level of terminal. This is also reasonable because substantially different flows pass through terminals at different levels, and it just doesn't seem economical to serve them equally frequently with the same tour – this was discussed in Sec. 4.7. Thus, the routes from any level-j terminal[1] will be assumed to serve all the level-(j-1) terminals in the level-j influence area, or the customers if j = 0 .

As a result, a set of influence areas and terminals (at all levels) defines the stops served from every terminal. Since the VRP solution defines the routes, the overall strategy is defined by a set of influence areas and a set of dispatching frequencies.

Because level-(n-1) influence areas are usually contained in much bigger level-n influence areas (otherwise terminals would not be cost-effective), the flow through a terminal usually is considerably smaller than the flow through the terminal feeding it. This, among other reasons such as restrictions to heavy vehicle travel on local streets, makes it economical to distribute items in loads smaller than those used to feed the terminal. Thus, each item-mile requires more vehicle-miles during distribution from the terminal than while being fed to the terminal. Consequently, in order to minimize vehicle-miles of travel, terminals should be centrally located within their influence areas. This is true for influence areas of all shapes.

The same location principle was applied to the one-dimensional terminal location problem described in Secs. 3.4 and 3.5. Although the optimal terminal locations obtained in Sec. 3.5 were not exactly in the center of each interval, the displacements were slight. Not surprisingly, the CA approximation with centered terminals was found to be quite accurate. Campbell's (1990) two-dimensional analysis confirms that this simplification leads to negligible errors; see also problem 5.1.

Unlike VRP zones, influence areas should *not* be elongated toward the depot; their shape should be selected to be as close to a circle centered around its terminal as possible, because this minimizes vehicle-miles. Of course, perfect circles cannot be used because they would not fill the space, but non-elongated shapes – "round" we call them – that approximate circles (e.g. squares, hexagons, and triangles) should be appropriate. The specific round shape used does not matter much (Newell, 1973).

It is thus possible to describe a near optimal system structure by the sizes of the various level influence areas, $I_j(\mathbf{x})$, as a function of position – together with the dispatching headways used at each level. As stated earlier, this reduces the very complex design problem to the determination of just a few decision variables. Building on this result, the following sections show how to estimate cost and develop a system design for various scenarios.

---

[1] We say that a terminal is of level-j if its items are transshipped a maximum of j times after passing through the terminal. The terminal serves a level-j influence area.

5.3 The One Transshipment Problem

We will focus first on the problem with only one transshipment (finding $I_0(\mathbf{x})$). This most common case is also useful as a building block toward multiple transshipment solutions. The one transshipment problem is similar to the classical facility sizing and location problem (Beckmann, 1968; Lösch, 1954 and Weber, 1929); it is slightly more complicated, however, because in addition to facility sizes, service schedules need to be determined. A recent review of the facilities location literature with an emphasis, as in this monograph, on models with few details is given in Erlenkotter (1988) – interestingly, a logistic cost function of the form $[\alpha x^a + \beta x^{-b}]$, as obtained in Daganzo and Newell (1986), is also proposed in this reference. Detailed models are reviewed in Love, et.al. (1988) and Brandeau and Chiu (1989).

In the spirit of the CA approach we will consider an imaginary subregion of $\mathbf{R}$ that is located r distance units away from the depot and exhibits a constant, stationary demand rate density (λ items per unit time and unit area) and a constant spatial customer density (δ customers per unit area). We will find the optimal dispatching frequency and the size of the influence area I^* in the imaginary subregion, assuming that vehicle routes are constructed as described in Chap. 4 – the subscript "0" is not used to index "I" because only level-0 influence areas are bcing considered in this section.

5.3.1 Terminal Costs

If no effort is made to coordinate the inbound and outbound schedules at a terminal, but the inbound and outbound headways (H^i ; H^o) are constant, the accumulation of items at the terminal for a specific destination is given by the vertical separation between step curves such as those of Fig. 4.9. The average inventory cost per item is then $(c_i/2)(H^i + H^o)$.

The maximum accumulation of items of any type cannot exceed the maximum vertical separation between the two curves. Since the item flow through the terminal is $D' = \lambda I$, the maximum vertical separation is $\lambda I(H^i + H^o)$. Thus, a conservative estimate for the holding costs per item at the terminal (the terminal serves an area of size I), is:

$$\left(\frac{c_i}{2} + c_r^t\right)\left[H^i\right] + \left(\frac{c_i}{2} + c_r^t\right)\left[H^0\right] + \left(c_i + c_r^t\right)H^t \qquad (5.1)$$

where H^t represents a (fixed) transfer time that an item must spend in the terminal even if H^i and H^o were zero, and c_r^t is the terminal rent cost coefficient (in monetary units per item-time).

This expression is important because it indicates that the waiting costs at the terminal are a sum of three separable components: a first term which only depends on H^i and is identical to the term that would have existed if the terminals had been the final destinations; a second term which only depends on H^o and is identical to the term that would exist if the terminal had been a depot producing items at a constant rate; and a third term which is a constant penalty.

For more realism we may also want to include a minimum rent to be paid per unit time, even if the maximum accumulation is zero, c_r^o . This will discourage the operation of very small terminals. Prorated to the items served in one time unit, the minimum rent is $c_r^o/(\lambda I)$; thus, the third term of (5.1) becomes:

$$\left(c_i + c_r^t\right)H^t + c_r^0/\left(\lambda I\right). \tag{5.2a}$$

This expression only accounts for the holding costs specific to the terminal; i.e. costs added by the transshipments, and not included in the sum of costs of distribution *to* the terminals (studied in Sec. 5.3.2, below) and the cost of distribution *from* the terminals (studied in Sec. 5.3.3).

In addition, items passing through the terminal must pay a handling cost penalty, which will have three terms: the cost of unloading the vehicle, the cost of sorting and transferring the items internally and the cost of loading the outbound vehicles. The first term is the same that would have to be paid if the terminal was a final destination, and the third term the same as if the terminal was the depot; these two terms will be captured in Secs. 5.3.2 and 5.3.3. The second term is terminal-specific. Its magnitude, on a daily basis, should grow roughly linearly with the number of items handled λI; expressed as a cost per item, it should be of the form:

$$c_f^o/\left(\lambda I\right) + c_f^t, \tag{5.2b}$$

where c_f^o and c_f^t are handling cost constants that depend on the nature of the items and the terminals. The total (motion plus holding) cost specific to the terminal is the sum of (5.2a) and (5.2b):

$$\textit{Terminal Cost per Item} \approx \alpha_5 + \alpha_6 / I \tag{5.3}$$

where $\alpha_5 = (c_f^t + c_i H^t + c_r^t H^t)$ and $\alpha_6 = (c_r^o + c_f^o)/\lambda$. Note that this expression is independent of H^i and H^o. It captures the costs not included in the sum of the costs of distributing to the terminals (inbound costs z^i), and the costs of delivering from the terminals (outbound costs z^o).

5.3.2 Inbound Costs

The total logistic cost, in addition to (5.3), must include all inbound and outbound costs; but these already have been studied.

The inbound cost would be given by the minimum of Eqs. (4.20) as applied to a problem where the terminals are the final destinations. Thus, v_{max} is the capacity of the vehicles used to feed the terminals, and the spatial density of customers δ becomes the density of terminals I^{-1} . Care must be exercised in solving the equations because, for large I, constraint (4.20c) may be binding; it may be optimal for vehicles to visit only one terminal at a time $(n_s^* = 1)$. Other constraints for route length or number of stops may also have to be considered (as explained in Chap. 4).

In solving the problem we may also want to alter the value of k (the VRP dimensionless constant for the distance added by each stop) to reflect the fact that stops will now be (roughly) on a lattice. Exercise 5.2 shows that this coefficient declines a little, but the change is only on the order of 15 percent. When there are more stops per tour than tours (this is highly unlikely when distributing to terminals) the change in k' is also small.

In any case, the minimum inbound cost will be a function of decision variable I only. This function will decrease with I because the more concentrated the demand becomes at fewer terminals $(I \rightarrow \infty)$ the cheaper it is to serve it. This was pointed out in Chapter 4. It should also be clear that the minimum cost per item can depend on parameters r and λ *but not on δ.* It will be denoted: $z^i(\lambda, r, I)$. The cost per unit area and per unit time, λz^i , will share the same properties.

5.3.3 Outbound Costs

The outbound cost per item depends on the density of destinations, but not on the distance from the depot. It can be calculated with the continuous approximation method, as if the terminal were producing items for the customers in its influence area, and averaging the result across the influence area in the usual way. Let $z_0(\lambda, r, \delta)$ denote the per-item cost of serving without transshipments a set of customers located r distance units away from a depot (the terminal) when the demand rate density is λ and the destination density is δ . This function is also given by the results in Chapter 4, but it may be somewhat different than for inbound costs because: (i) customers may be randomly scattered (not on a lattice like the terminals), (ii) vehicles may have smaller capacities, (iii) travel speeds may be lower, and (iv) perhaps all the customers do not need to be visited with each dispatch (recall the discussion of Sec. 4.6).

According to the continuous approximation approach, the cost per item delivered from the terminal can be approximated by averaging $z_0(\lambda, r, \delta)$ over r , where r is now the distance from points in the influence area to its terminal (recall the discussion of Eq. (3.11c) in Sec. 3.3.4). We will denote this average, independent of r but a function of I , by a capital "Z" superscripted by "zero" – the level of the influence area – Z^0 . Thus:

$$Z^0(\lambda, \delta, I) \cong E_r[z_0(\lambda, r, \delta)].$$

We saw in Chapter 4 that z_0 increased with r, and that in some cases (e.g. when the vehicles are filled to capacity – see Eqs. (4.10) and (4.21)) it does so linearly. It is thus reasonable to substitute $E_r[z_0(\lambda, r, \delta)]$ by $z_0(\lambda, E[r], \delta)$, and to approximate E(r) by a simple function of I. Since influence areas will be drawn to approximate circles and the density of destinations is approximately uniform, we can assume that E(r) is 2/3 the maximum distance from the terminal, $(I/\pi)^{1/2}$, and thus:

$$Z^0(\lambda, \delta, I) \cong z_0\left(\lambda, \frac{2}{3}\sqrt{\frac{I}{\pi}}, \delta\right) = z_0\left(\lambda, 0.38 I^{1/2}, \delta\right), \tag{5.4}$$

which increases with I , (linearly with $I^{1/2}$ in some important cases). We are now ready to see how the system can be designed.

5.3.4   The Design Problem

The next step consists in writing a logistic cost function that relates the total cost per item distributed to the decision variables of the problem. In our particular case, the total cost per item distributed is the sum of the terminal, inbound and outbound costs:

$$\underset{terminal}{(\alpha_5 + \alpha_6/I)} + \underset{inbound}{z^i(\lambda, r, I)} + \underset{outbound}{Z^0(\lambda, \delta, I)} \tag{5.5}$$

Figure 5.4 depicts these functions and their sum.

The value of I , I^* , that minimizes this expression is the size of the influence area which we would like to use. Values of I larger than the service region size, $|\mathbf{R}|$, do not need to be considered. The optimum influence area size, I^*, should usually grow with the distance from the depot but it can also be independent of r, e.g., as occurs with the "cheap item" scenario leading to Eq. (4.21). This will be illustrated with the example of Sec. 5.3.5.

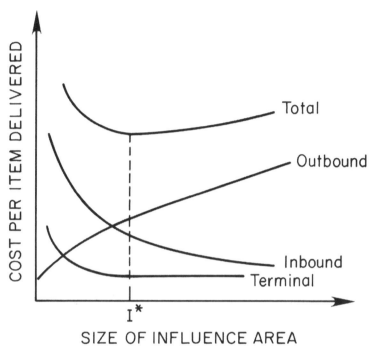

Fig. 5.4   Logistic cost components and influence area size

The minimum cost obtained with the above expression, denoted $z_1^*(\lambda,r,\delta)$ because one transshipment is used, should be compared to the cost of distribution without transshipments, $z_0(\lambda,r,\delta)$. Only if $z_1^* < z_0$ should transshipments be used. The cost per item with *upto* one transshipment, z_1, is the minimum of z_1^* and z_0: $z_1 = \min\{z_0, z_1^*\}$.

Figure 5.5 depicts this relationship as a function of r for constant λ and δ. As we have indicated, z_0 increases with r; z_1^* also increases with r, but at a lower rate for large r. If the curves don't intersect, then terminals don't have the potential for reducing cost. We have already seen that terminals are beneficial (see Fig. 5.1) if there are restrictions to the size of a local delivery vehicle and/or route length limitations, but in the absence of such limitations transshipments are likely to be unnecessary (see Daganzo, 1988, and the discussion at the beginning of Chap. 6).

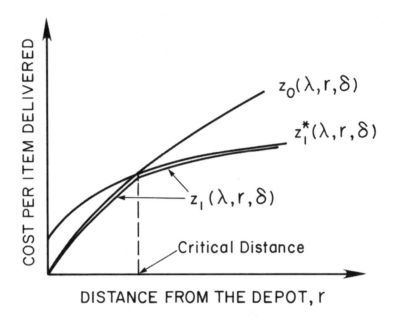

Fig. 5.5   The cost of serving a point r distance units away. (Adapted from: Daganzo and Newell, 1986)

The expected total cost per unit time over any subregion of **R** , **P** , can be obtained even before a solution scheme is constructed, by integrating $\lambda z_1(\lambda,r,\delta)$ over **P** . Expressed per unit time, the total cost, again denoted by a capital "Z" , is:

$$Z_T^l(P) \cong \int_P \lambda\, z_l\left(\lambda,r,\delta\right)dx ,$$

where λ , r and δ can be slow varying functions of **x** . The subscript "T" alludes to "total cost per unit time" and the superscript to the maximum number of transshipments allowed.

To illustrate how a design can be obtained, Fig. 5.6a depicts the loci of points in **R** for which level-0 influence areas have five different sizes. This could be the result of solving the idealized model for different points in **R** , with different λ , r and δ . These sizes were chosen to increase relatively fast to make the partitioning more difficult. Points in between the curves require intermediate sizes. Figure 5.6b shows a possible partition of **R** that conforms fairly well with the stated requirements.

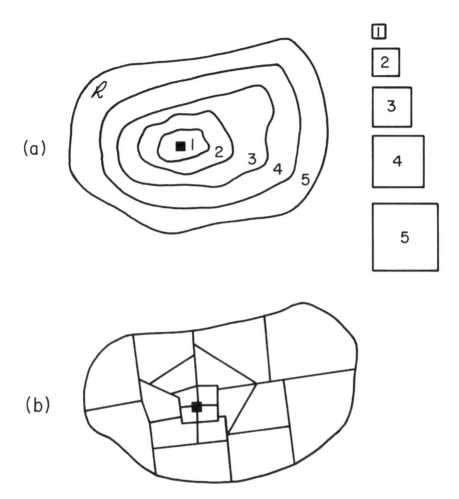

Fig. 5.6.   Implementation example. (Source: Daganzo and Newell, 1986)

In general, the complete design can be obtained as follows. First carve out "round" influence areas that pack and conform to the calculated sizes $I(\mathbf{x})$ as well as possible, as we have just shown. Then locate the terminals near their middle, obeying any local constraints that may exist. Finally, determine with existing methods (see Sec. 4.2) the optimal operating strategy within each influence area, separately from the others.

Note from the figure that while many points in $\mathbf{R}$ do not belong to an influence area of the right size, few have to be enclosed in areas that are off by more than 50 percent from the target size. Larger discrepancies should be rare in practice. Discrepancies of typical magnitude introduce little error into the resulting cost, $Z_T{}^1(\mathbf{R})$, since the logistic cost function (5.5) is usually rather flat around its minimum with respect to I – see Fig. 5.4. The solution to problem 3.10 illustrates this fact by examining cost functions (5.5) of the common form: $\alpha I^a + \beta I^{-b}$ (a,b ≤ 1). For this kind of expression the chosen value of I can depart from the optimum by as much as 50 percent, and the resulting cost will still be within a few percent of the optimum. When a and b are smaller than 1 the solution is even more robust than the EOQ expression (the case with a = b = 1). We can be reasonably sure as a result that demand points do not have to be enclosed in influence areas of the precise size for a solution to be near-optimal.

For example, if (i) moderately priced goods have to be delivered to fixed retail outlets, (ii) vehicles can make multiple stops, and (iii) no terminal economies of scale exist ($\alpha_6 = 0$), then Eq. (5.5) consists of a constant, a term proportional to $I^{1/2}$ and a term proportional to $I^{-1/4}$. Then, I could be 1.5 times larger or smaller than I^* and cost would only increase by about one percent. Although not quite so robust, the example about to be introduced exhibits a similar behavior. Among those problems explored by the author (involving various underlying metrics, deliveries of people and goods, routes with and without multiple stops, deliveries to fixed retail outlets, and individually located customers, etc. ..), the example corresponds to the set of conditions that makes the cost most sensitive to I .

5.3.5 *Example*

Here we consider a region **R** with constant λ and δ . Line-haul vehicles shuttle between a distribution center and consolidation terminals. Neither local nor line-haul vehicles are allowed to make multiple stops because the cost (and delay) of a stop is large compared with that of the moving portion of the trip. This could happen for air transportation of valuable goods. Campbell (1990c and 1993) contains a similar analysis of problems with multiple stops. In our case, local transportation vehicles pick up their loads at the consolidation terminals and distribute them (non-stop) to destinations scattered over the terminals' influence areas. Local vehicles are assumed to have a small capacity, v_{max} , and to travel full; i.e, the solution to (4.20) is $n_s = 1$ and $v = v_{max}$. To make things easier we also assume that the pipeline inventory cost and rent costs can be neglected; i.e., $c_h = c_i$. We then see that the minimum cost is of the form:[2]

$$z_0(\lambda, r, \delta) = \frac{\alpha_1 + \alpha_2}{v_{max}} + \alpha_4 v_{max} = constant + \frac{2rc_d}{v_{max}} + \frac{c_h \delta}{\lambda} v_{max} .$$

---

[2] When $n_s = 1$ the local distance vanishes. The average customer demand rate is $D' = \lambda/\delta$.

To simplify the notation, we will ignore the constant term and introduce two constants "a" and "b" ($a = c_h\delta$ and $b = 2c_d/2.7$) so that:

$$z_0(\lambda,r,\delta) = \frac{av_{max}}{\lambda} + \left[\frac{2.7b}{v_{max}}\right]r. \tag{5.6a}$$

The first term is the stationary holding cost and the second term, the component of transportation cost that is sensitive to distance. For this example, z_0 is independent of δ, and so is the outbound cost function (5.4):

$$Z^0(\lambda,\delta,I) \cong \frac{a\,v_{max}}{\lambda} + \frac{b}{v_{max}}I^{1/2}. \tag{5.6b}$$

Inbound transportation to the terminals is assumed to take place on larger vehicles, of capacity $v'_{max} > v_{max}$ and cost per mile c'_d , operated at capacity so that the cost $z^i(\lambda,r,I)$ will be (for a demand rate $D' = \lambda I$):

$$z^i(\lambda,r,I) = \frac{\alpha_1 + \alpha_2}{v'_{max}} + \alpha_4\,v'_{max} = constant + \frac{2rc'_d}{v'_{max}} + \frac{c'_h}{\lambda I}v'_{max}.$$

Again, using $a' = c'_h$, $b' = 2c'_d$ and ignoring the constant we can write:

$$z^i(\lambda,r,I) = \left(\frac{a'v'_{max}}{\lambda}\right)\frac{1}{I} + \left(\frac{b'r}{v'_{max}}\right). \tag{5.6c}$$

The first term of this expression represents inventory cost, and the second the cost of overcoming distance. Inventory cost must increase with the number of destinations; as such it is proportional to I^{-1} . Other costs (handling, etc.) that don't depend on I , r , or λ would appear as part of the omitted additive constant.

Let us assume that terminal costs are proportional to flow ($\alpha_6 = 0$) . Then they can be ignored, and the optimal influence area size is the result of a trade-off between the cost of overcoming outbound distance from the terminals (2nd term of (5.6b)) and the stationary inventory cost from inbound distribution (first term of (5.6c)); the solution is:

$$I^* \cong \left[\frac{2\, a'\, v_{\max} v'_{\max}}{b\lambda} \right]^{2/3}. \tag{5.6d}$$

Therefore the one-transshipment cost is:

$$z_1^* \cong a\, \frac{v_{\max}}{\lambda} + \frac{b'r}{v'_{\max}} + 1.89 \left(\frac{b}{v_{\max}} \right)^{2/3} \left(\frac{a'}{\lambda} v'_{\max} \right)^{1/3}. \tag{5.7a}$$

The optimal size of the influence area increases with the 2/3 power of the vehicle capacities and decreases with the 2/3 power of the demand density; it does not depend on the distance, r , from the distribution center. This is logical, because changing r does not alter the terms traded off. These qualitative conclusions, however, are specific to the conditions of the example.

To see how they would change, assume that the inbound vehicles, still restricted to making one stop, now can carry as many items as desired $(v'_{\max} = \infty)$. Then, the loads carried would be the result of an EOQ trade-off, and instead of (5.6c) we would have:

$$z^i(\lambda, r, I) = 2 \left(\frac{a'b'r}{\lambda I} \right)^{1/2} \tag{5.6e}$$

and

$$I^* \cong \left(\frac{2\, v_{\max}}{b} \right) \left(\frac{a'b'r}{\lambda} \right)^{1/2}. \tag{5.6f}$$

The optimal solution is no longer insensitive to r; it grows with r as indicated earlier. It also varies with a smaller power of λ and a larger power of $v_{\max}$. The optimal cost also depends on r and λ , although somewhat differently:

$$z_1^* \cong a\, \frac{v_{\max}}{\lambda} + 2.83 \left(\frac{b}{v_{\max}} \right)^{1/2} \left(\frac{a'b'r}{\lambda} \right)^{1/4}. \tag{5.7b}$$

It should be easy to design a system with influence areas close to I^* for most points. Failure to select an I equal to I* does not result in large increases in cost. For both examples a 30 percent deviation from I* results in a cost increase below three percent; for 20 percent deviations cost increases less than one percent. These percentages refer only to the two cost terms that depend on I ; otherwise, the percentages would be even smaller. The dependence of cost on I (and its sensitivity to errors in λ and δ) tends to weaken even more when multiple stops are allowed; the conditions of the example are unfavorable.

5.4 Refinements and Extensions

This section, which may be skipped on a first reading, addresses the following subjects which are extensions to the simple model of Sec. 5.3: (i) synchronization of the inbound and outbound transportation schedules to reduce terminal holding costs; (ii) treatment of location/routing constraints cutting across distribution levels; (iii) consideration of time-varying demand, with and without uncertainty; and (iv) development of discriminating strategies when conditions warrant.

The analysis of Sec. 5.3 was possible because inbound and outbound vehicle routes and schedules from the terminal could be set independently of each other. This decomposition allowed the results of Chapter 4 to be invoked, yielding simple inbound and outbound cost expressions. Because some of the extensions explored in this section link the inbound and outbound operations, a conditional decomposition method is used repeatedly. It entails the identification of suitable decision variables, conditional on which the problem decomposes across levels. We recommend a similar approach whenever inbound and outbound operations are coupled.

5.4.1 Schedule Coordination

It was assumed in Sec. 5.3 that inbound and outbound operations were independent. Yet, terminal holding costs can be reduced through synchronization. (Chapter 4 showed how synchronization of transportation and production schedules could reduce holding costs; something similar happens here).

If we restrict the inbound headway to the terminal H^i to be an integer multiple of the outbound headway H^o , or the other way around, then it is possible to synchronize the arrivals and departures as shown in Fig. 4.10. This synchronization allows the average time in the terminal to be reduced by the smaller of the two headways: min $\{H^i ; H^o\}$. It is as if the departure curve on Fig. 4.9 had been shifted to the left by an amount, H , resulting in the pattern of Fig. 4.10. Then, the maximum accumulation is reduced by $\lambda I \min\{H^i , H^o\}$, and the terminal cost per item (5.3) becomes:

$$\alpha_5 + \alpha_6/I - \min\{H^i \; ; H^o\}[c_i + c_r],$$ (5.8)

which no longer is independent of H^i and H^o .

Conditional decomposition: The method we are about to present works even if the outbound headways from the terminal are not equal for all the delivery districts; but we shall assume for the moment that they are. The total cost will be expressed as a function of I , H^i and H^o . Conditional on these three variables (instead of only one, I), the total logistic cost per item decomposes in three independent components: (i) an inbound motion cost, z_m^i ; (ii) an outbound motion cost, z_m^o ; and (iii) the terminal costs plus all holding costs. Thus, the new logistic cost function is expressed as follows:

$$z_m^i\left(\lambda, r, I, H^i\right) + z_m^o\left(\lambda, \delta, I, H^o\right) + \left(c_i + c_r\right)\max\left[H^o; H^i\right] + \left(\alpha_5 + \alpha_6 I^{-1}\right),$$ (5.9)

where we have assumed that the rent costs only need to be considered for the terminal. If rent costs at the origin and the destinations cannot be neglected then a term of the form $c_r(H^i + H^o)$ should be added to (5.9). In this case, the headway choices may differ from those recommended below. In both cases, however, the choices arise from the minimization of a simple logistic cost function of three variables: I, H^i and H^o.

The inbound motion cost term in (5.9) assumes that the inbound routes have been optimized for the given set of terminals and inbound headways, independently of all outbound decisions. This cost can be estimated by the minimum of the first three terms of Eqs. (4.20) with respect to n_s for a given v since the delivery lot size to a terminal, v , is fixed by H^i: $v = \lambda I H^i$. We are pretending here, as in Sec. 5.3, that the terminals are the final destinations. The outbound motion cost for delivery to the customers can be obtained in a similar way, also conditional on the delivery lot size to the customers, $v = (\lambda/\delta)H^o$.

For most problems the inbound and outbound cost per item, z_m^i and z_m^o , will be decreasing (or non-increasing) functions of H^i and H^o respectively. This is logical since with longer headways more goods will have accumulated with every dispatch and they can be distributed more efficiently.[3] The

---

[3] This assumes that vehicles make many stops and therefore it is only true up to a point. Once H^i (or H^o) is so large that each destination requires a full vehicle load on each visit, increasing H is no longer beneficial.

reader may wish to verify this analytically from Eqs. (4.20), with and without other complications, as discussed in Chapter 4.

It follows from these properties that the least cost given by (5.9) is achieved when $H^i = H^o$; it should be clear that if the smaller of the two headways is not equal to the other, increasing the smaller one until it equals the largest will reduce cost: clearly, the holding cost does not change, and we have already said that the motion cost declines with an increasing H. Thus, we let $H^i = H^o = H$, so that (5.9) becomes:

$$z_m^{\ i}(\lambda, r, I, H) + z_m^{\ o}(\lambda, \delta, I, H) + (c_i + c_r)H + (\alpha_5 + \alpha_6\, I^{-1}),$$

whose minimum (I^*, H^*) is the desired solution.

To find it we can hold I constant and minimize the first three terms, the inbound plus outbound costs, with respect to H; the result is of the form:

$$z_m^{i*}(\lambda, r, I) + z_m^{o*}(\lambda, \delta, I) + \\ + (c_i + c_r)H^*(I) + (\alpha_5 + \alpha_6\, I^{-1}),$$ (5.10)

where $H^*(I)$ is the optimum headway for a given influence area size. This expression can now be treated like (5.5) for design purposes.

Schedule synchronization takes some effort and may add to the total cost because the operation of the system is more complex. Obviously, it should only be used if the gains outweigh the complexity penalty. The higher the time value of the items the larger the gain and the more desirable synchronization becomes. As an exercise, the reader may want to repeat the simple example presented in Sec. 5.3.5, assuming that the schedules are synchronized. (Note that in this example vehicles make a single stop, and therefore we cannot assume that $H^o = H^i$; see problem 5.3).

Different outbound headways: Suitably modified, the approach we have described can be applied when the influence area is not homogeneous; e.g. if λ and δ change within the influence area.

In this case, as illustrated in Fig. 5.7, the vehicle routes to parts of the influence area with different characteristics (or sectors, j) ought to have different numbers of stops and different headways, H^o_j. As in Chapter 4, these sectors should be no smaller than the districts covered by one vehicle. Of course, this restriction is irrelevant if vehicles make only one stop.

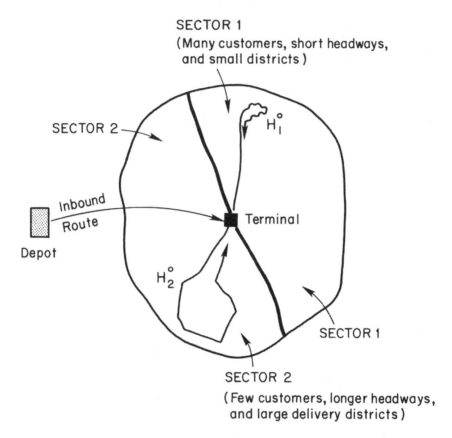

SECTOR 1
(Many customers, short headways,
and small districts)

SECTOR 2

Inbound
Route

Depot

H_1^o

Terminal

H_2^o

SECTOR 1

SECTOR 2
(Few customers, longer headways,
and large delivery districts)

Fig. 5.7.   Distribution within an influence area with two sectors

The decision variables are I , H^i and $\{H_j^o\}$. If the outbound headways are multiples or submultiples of H^i , then Eq. (5.9) holds with the following modifications: (i) the outbound motion cost is the demand weighted average of the costs of each sector, $z_{m,j}^o (\lambda,\delta,I,H_j^o)$, and (ii) the waiting cost is the demand weighted average of $(c_i + c_r)\max\{H^i , H_j^o\}$.

As before the $z_{m,j}^o$ are decreasing functions of H_j^o , and thus *outbound headways should be no smaller than the inbound headway*. (Blumenfeld et al. 1986, and Daganzo, 1990, have addressed this issue.) Most likely, and even if the demand is quite heterogeneous within I, the solution with $H_j^o = H^i = H$ will be close to optimal, and we can obtain $H^*(I)$ in the same manner as we described.

A more accurate approach described in the above mentioned references would find first the optimal $\{H_j^o\}$ *conditional on I and H^i* . This is easy because each H^o_j can be obtained independently of the others as the result of a trade-off between the outbound transport cost in its sector *alone*, and the corresponding waiting cost: [4] $(c_i + c_r)\max\{H^i, H^o_j\}$ One could then find the optimal H^i and I , either numerically or analytically.

Further reductions to holding cost: In the discussion of dispatching strategies it was assumed, even when the schedules were coordinated, that the space needs at the terminal were the sum of the space needs for all the outbound destinations from the terminal. This assumption is conservative because it ignores that the need for storage space can be reduced if one staggers the delivery schedules, as was discussed at the end of Sec. 4.3. If vehicles depart the terminal once every two days for points in the influence area, we implicitly assumed that the maximum accumulation occurred at the same time for all outbound routes. But if some of the routes (1/2 of them, say) depart on even days and the other half depart on odd days, then the maximum accumulation will be reduced. It will be more difficult, though, to coordinate the deliveries to the terminal with the improved staggered schedule from the terminal. Obviously, the advisability of staggering outbound dispatches will depend on the specific situation. In any case, the methods we have introduced in connection with Eqs. (5.5) and (5.10) still apply.

5.4.2 Constrained Design

Here we address two types of design restrictions, which influence the solution approach in different ways: (i) constraints to individual decision variables, and (ii) constraints to sets of variables. The first type of restriction arises in connection with tactical problems, which are used as an illustration for the solution approach to (i).

Tactical problems: If some of the decision variables are fixed, the optimization process is often simplified. This situation is common for short term problems, where the terminal locations are given, but the vehicle schedules and routes need to be determined.

If so few terminals are available that all should be used, then each influence area will be greater than ideal. One could then easily carve the region into influence areas around each terminal, perhaps allocating every customer to the nearest terminal. Ideally, one would like to allocate customers using a marginal cost rule, ensuring that customers near the boundaries are

---

[4] It is shown in Daganzo (1990) that, even if the H^o_j are not restricted to be integer multiples of H^i, the optimal solution is an integer multiple of H^i.

not better off in the neighboring influence area, but this is quite laborious and unlikely to change the size of the influence areas enough to matter for cost calculations. (A marginal allocation is reasonable because outbound distribution costs per unit area $\lambda z_0(\lambda, r, \delta)$ increase with r and as a result the total cost is convex in the zone dimensions). Once the partition has been completed, a cost estimate − as well as the optimal headways and routes − can be obtained for each influence area by minimizing an appropriate logistic cost function, e.g. (5.5) or (5.9), with a fixed I. Note that no minimization is necessary with Eq.(5.5). Finally, the detailed solution can be fine-tuned by computer by testing whether marginal customers near the boundaries should switch terminals.

If there are so many terminals that we don't know beforehand which ones should be operated, a preliminary step should be taken to make this decision. Based on the given arrangement of terminals, we would define a minimum feasible influence area size, $I_{min}(\mathbf{x})$, as a function of position. We would then obtain for different values of $\mathbf{x}$ an ideal influence area size, $I^*(\mathbf{x})$, by minimizing (5.5) or (5.9) subject to $I(\mathbf{x}) \geq I_{min}(\mathbf{x})$; the result would then be used to decide which terminals to operate. Of course, if there are considerably more terminals than needed this constraint plays no role. Then, the terminals can be selected based on the solution to the strategic problem.

In the short term we may also have to account for restrictions in the flow through some of the terminals, which essentially impose a limit on the size of their influence areas. If such flow restrictions can be translated into an upper bound restriction to $I(\mathbf{x})$, $I(\mathbf{x}) \leq I_{max}(\mathbf{x})$, then the preliminary step can still be carried out as indicated. The detailed allocation of customers to terminals during fine-tuning, however, must recognize the existence of the flow restrictions. The optimal allocation can be obtained by linear programming. (If one thinks of terminals as sending a flow equal to their capacity to destinations requesting a flow equal to their demand, and we include an extra destination to which the slack capacity is sent at zero cost, then the flow allocation problem reduces to the Hitchcock transportation problem of linear programming). For our problem the costs have a special structure that relates to the geographical distribution of customers; i.e., customers that are close geographically have similar costs from all the terminals. As a result, it is not difficult to prove that the set of customers to be served from any terminal should be a well defined region around the terminal. Thus, if the terminal capacities change, only the boundaries to the regions should move. Hall (1989a) has discussed very cogently the geometric interpretation of the Hitchcock problem.

Multilevel constraints: The problem just discussed was viewed as a design problem with a constraint on the size of the influence areas. Constraints affecting only inbound, or only outbound, logistic operations are also easy to incorporate; one simply needs to make sure that the expressions for inbound and outbound motion costs, z^i_m and z^o_m, properly reflect the effect of the constraints. We have already seen how to develop these expressions under a variety of conditions in Chap. 4.

Constraints that cut across levels are a different matter. This occurs, for example, if there is a maximum time allowed for an item between the origin and the final destination, or a limited transportation budget and/or fleet size for both distribution levels. Multilevel constraints like these, can be captured with an extra level of decomposition. In addition to I, H^i and H^o, one should include one or more conditioning variables that will decompose the logistic cost function into independent subcomponents.

For an example in which total time is limited to an amount t^{max}, e.g. for the distribution of perishable items, we could use maximum times for the inbound and outbound operation, $t_{(1)}{}^{max}$ and $t_{(2)}{}^{max}$, and write the equivalent of Eq. (5.9) also as a function of $t_{(1)}{}^{max}$ and $t_{(2)}{}^{max}$. The resulting 5-variable logistic cost function can be minimized subject to constraints on $t_{(1)}{}^{max}$ and $t_{(2)}{}^{max}$ that will ensure t^{max} will not be exceeded (e.g. $t_{(1)}{}^{max} + t_{(2)}{}^{max} + \max (H^i, H^o) + H^t \leq t^{max}$). A solution to this problem, for a newspaper delivery network, is discussed in Han (1984) and Han and Daganzo (1988). For this problem, in contrast to most other applications, as one moves farther away from the depot both the size of the influence areas and the length of the delivery routes *decline*.

If there is a total transportation budget or a fleet size constraint the same method can be used; one would allocate budgets/fleets to the two distribution levels and solve the resulting 5-variable problem. This technique can be applied to design elevator systems with skylobbies (see problem 5.5).

5.4.3 *Variable Demand*

This subsection discusses stochastic and deterministic demand variations.

Stochastic demand: Here we examine the implications of random (unpredictable) demand at the destinations. Random demand requires extra inventories at the destinations – in the form of safety stocks as seen in Sec. 2.5 – and also at the terminals.

We will first see that the decision to hold a certain safety stock at a destination can be separated from the routing decisions, conditional on the inbound and outbound terminal headways. A similar decomposition had been already introduced in Sec. 2.5 for one-to-one distribution problems.

We will then examine the need for inventories at the terminals (warehouses).

As explained in Chapter 2, we assume that the safety stock carried by a customer (destination) depends on the time between deliveries and requests. With deliveries to many customers, however, it is unreasonable to assume that one would dispatch on request, at the precise time when the customer request arrives, lots of specific sizes as assumed in Chapter 2; it would then be impossible to construct "peddling" delivery tours. Rather, one would attempt to coordinate deliveries to all the customers by establishing a dispatching schedule from the terminal at headways H^o—a decision variable – and allowing customers to decide whether or not they desire a delivery on any given dispatch as well as the size of the delivery, $v >$ $D'H^o$. Note that the fixed lead time model of Chapter 2 arises with $H^o = 0$.

For this operating scheme the inventory accumulation at a destination depends on the fixed lead time and on H^o, *but it is independent of the transportation routing decisions.* The solution to problem 5.6 reveals that the holding cost per item includes: a term proportional to v, $c_h v/D'$, that represents the load make-up cost as in Chap. 2, a safety stock component that increases slightly with H^o but is independent of v as in Chap. 2, and a new term, $c_h H^o$, that captures the discreteness of the transportation schedule. This is discussed in more detail in Daganzo and Newell (1987).

Then, conditional on I, H^i and H^o, the customers' decisions about v are independent of the routing decisions; the problem decomposes. Transportation costs decrease with H^o, while holding costs increase. To enforce rational customer behavior, e.g. discouraging small orders, we will pretend that an amount c_p is charged to the customer for each delivery. Whether real or fictitious (if the customers are part of the same firm) this charge can be used to control the customer lot sizes and in the process achieve some overall goal such as maximizing profit, minimizing the sum of costs to the supplier and the customers, etc... We show below how the various logistic cost components can be expressed as functions of I, H^o, H^i and c_p.

Aside from additive terms independent of v, the motion cost per item paid by a customer will be c_p/v and the holding cost will be $(c_h/D')v$. The optimal lot size chosen by the customer is thus the result of an EOQ trade-off between those two costs: $v^* = [c_p D'/c_h]^{1/2}$, provided v^* is greater than $D'H^o$; it is $D'H^o$ otherwise. Such a customer would place an order on one out of every $v^*/(D'H^o)$ dispatches, on average. If all customers were roughly alike, the reciprocal of this ratio would also represent the fraction of customers requesting service. The effective density of delivery stops δ_e in the region would then be:

$$\delta_e = \delta[\min\{1, D'H^o[c_p D'/c_h]^{-1/2}\}] .$$

The average cost paid by a customer per item delivered is also the result of the EOQ trade-off:

$$\begin{pmatrix} customer \\ cost \\ per\ item \end{pmatrix} = 2[c_p c_h /D']^{1/2} + c_h H^o + CONST. \qquad , if\ v^* \ge D'H^o$$

$$= c_p /(D'H^o) + c_h H^o + c_h H^o + CONST. , otherwise.$$

Note that this expression is an increasing function of c_p and H^o . In other cases (e.g. with different customers or other reorder strategies) the lot size, stop density and cost relationships would be similar; in particular, the effective stop density would still be a non-decreasing/non-increasing function of H^o and c_p .

This effective density, together with H^o and I determines the supplier's outbound motion costs from the terminal, z^o_m , as discussed in Sec. 5.4.1. Inbound motion costs are also determined as in Sec. 5.4.1, from H^i and I . Both the inbound and outbound motion costs are insensitive to fluctuations in the number of items dispatched from the terminal every H^o because these costs only depend on the average terminal throughput λI (see the discussion pertaining to Eq.(4.7) in Sec. 4.3).

The supplier's holding costs at the terminal would increase with the fluctuations in the number of items demanded per dispatching headway. Of order $[\lambda I H^o]^{1/2}$, however, the fluctuations should be small compared with the mean $[\lambda I H^o]$ because influence areas contain many customers. Thus, the added warehousing costs should be small compared with the deterministic holding costs, which in turn are small compared with the motion costs. As a result warehousing costs can be neglected as a first approximation, and terminal holding costs can be approximated by a simple function of H^i, H^o , and I , as described earlier in this chapter.

The sum of the inbound, outbound and terminal costs captures the supplier's logistics cost per item as a simple function of c_p , I , H^o and H^i . It is then a simple matter to obtain the c_p , I , H^o and H^i that minimize any desired combination of the supplier and customer costs.

If desired, the analysis can be refined by including warehousing costs at the terminal. These costs are likely to be significant only if the demand is highly variable and unpredictable, and order response time is critical. Although, there is an extensive body of literature on inventory control for a hierarchical system of warehouses (see Schwartz, 1981, and Anily and Federgruen, 1993, for example), very simple models should suffice for our purposes; sensible decisions can be reached without resorting to very detailed models. Problem 5.7, an illustration, describes a situation where warehousing costs must be traded-off against transportation costs. In the problem, order response time is so critical that individual items are delivered immediately upon request from warehouses by a very expensive and expeditious transportation mode. Cheap transportation is used to feed the warehouses. The influence area size is the key decision variable for this problem because warehousing costs decrease with I (with larger I the fluctuations in throughput are smaller) but the number of expensive vehicle-miles traveled increases with I (since the distance traveled per item is proportional to $I^{1/2}$).

When the ratio of inventory cost to local distribution cost is sufficiently high warehouses should pool risk by operating in clusters that can share inventory by coordinating their local distribution. This has the potential to reduce cost even more. Two different cooperation methods are possible: (i) periodic redistribution of goods as per a transportation problem of linear programming (TLP); or (ii) continual re-balancing by serving customers roughly equidistant from two warehouses by the one with the least inventory. The automobile industry essentially uses an extreme version of (ii) - since automobile purchasers are usually served by the nearest dealer having the desired car. Hybrids of the two strategies can also be used. Hierarchical schemes, where a lead warehouse holds extra inventory for potential redistribution, can be used as well, but they are less efficient than (i). Strategy (i) should be optimized by treating the influence area size, cluster size, safety stock level and re-balancing period as decision variables. In case (ii), the re-balancing period is not an issue. The objective function for scheme (i) is easy to write using the expected distance formula for the stochastic TLP given in Daganzo and Smilowitz (2004). The objective function for (ii) is more complicated - since the safety stock level affects the frequency deliveries from the second or third-nearest warehouse in a nontrivial way - but this too can be done; see Daganzo and Erera (1999).

Hopefully, this brief discussion will have convinced the reader that it is possible to capture stochastic effects with minor modifications to the approach we have described.

Non-stationary demand: At the tactical level, i.e. when only vehicle routes and schedules can be changed, non-stationary conditions do not introduce major difficulties. If the average demand rate $\lambda(t, \mathbf{x})$, the customer density $\delta(t, \mathbf{x})$ and the given set of terminals vary slowly with time the CA approach can be used. First we divide the time line into intervals with quasi constant conditions. Then, we find the optimal customer allocations, vehicle routes and frequencies for each interval independently as if the number of terminals, λ and δ, didn't change with t . (The chosen solution and the cost estimated for each interval should recognize stochastic effects if they are deemed important. For this to be possible, however, the system should nearly reach a steady state in each interval.)

Each solution is then adopted for its time interval. The average cost over time can be approximated by the weighed average of the costs for the intervals.

The strategic problem, including the number and location of the terminals can be addressed in a similar way if terminals can be opened, closed and relocated with little cost. If this is not the case the problem is considerably more complicated because small changes in I from one time interval to the next, as would result from the CA approach, might require that most terminals be relocated, and the relocation cost is hard to define.

If relocation is expensive we would like to relocate few terminals if I changes little. In fact, we would like to change only the absolute minimum number of terminals; i.e. "x" percent of them if I changes by "x" percent. This can be achieved if terminals are located approximately on a square lattice, and in every time interval the influence area size is restricted to take a value from the set $\{2^{K}I_{0}\}$ for some integer K . (Note that if terminals are located on a square lattice, one can obtain another square lattice with twice the I oriented at $45°$ with the old, by eliminating 1/2 of the terminals.) With the I restricted in this manner, it is a simple matter to define relocation costs as a function of the change in I from interval to interval. Conditional on I , then, the remaining costs can be obtained as the solution to the tactical problem. Thus, it should be possible to use a dynamic programming formulation to determine the best sequence of I's for a set of consecutive time intervals, where the dynamic programming stage is the time interval and the state is the I for the current interval.

Alternatively, one can use a human/computer hybrid method. The human would specify a changing pattern of terminals over time and the computer would readily solve the tactical problem, returning the total cost including the cost of terminal relocations.

More sophisticated methods don't seem necessary because long term forecasts as would be needed for strategic analysis are not likely to be reliable. Perhaps we should take a cue from nature in seeing how a logistic structure should adapt to changes in its environment without a forecast for future conditions. As a tree grows taller, new branches overshadow old branches, which may atrophy and die, but the larger older established branches survive. Because of the "cost" of growing new branches (opening a terminal), the tree does not totally redesign itself with each change in the environment; rather it preserves a large portion of its structure and builds on it. Moreover, the tree adapts to the future without knowing it – at best it uses the recent past experience as an indication of things to come. In light of this, and since the logistic cost is not very sensitive to the specific location and number of terminals, it should be possible to respond to a change in demand by opening and closing only a small fraction of the total number of terminals, and still obtain a configuration that will yield near minimal cost for the new conditions and the anticipated immediate future. Campbell (1990a) explored a few heuristic strategies for updating terminal locations under some scenarios and his conclusions agree with the above.

5.4.4 Discriminating Strategies

So far in this chapter we assumed that customers in the same general area received the same type of service in terms of delivery frequency and type of vehicle route. No attempt was made to discriminate across customers based on their individual characteristics. We had seen in chapter 4 that if some customers are much larger than others, or request substantially different items, it might be cost-effective to treat them differently, perhaps even serving them with different transportation systems. The same phenomenon can be expected of systems with terminals but the question is now whether or not all customers should be served through the terminals; large ones may be better off with direct service.

Approach: The conditional decomposition method introduced in this section can also be used to explore this possibility. Given a set of terminals, the tactical problem could be solved by dividing the set of customers into a set that is served through the terminals and another set which is served without a transshipment, organizing the distribution process for the two sets separately, calculating the cost, and then comparing the results for different customer partitions.

For the decomposition based on customer partitions to be successful one needs to focus only on a few partitions that have a chance of being optimal because the number of arbitrary partitions can be astronomical. Depending on the specific situation at hand, a set of candidate partitions should not be difficult to identify based on physical considerations. (The reader may wish to solve problem 5.8 as an exercise.)

For example, if as in Sec. 4.7 the items have different values and the customers have different sizes, one may prove that the destinations with the largest dollar demand per unit time should be served direct and the rest through the terminal. This happens because one can reduce the holding cost by swapping customers between the two shipping methods without changing the transport routes and cost.[5] This property of the problem allows us to use the fraction of customers that are served without transshipments, f_o , to define the two customer classes and decompose the problem. The f_o leading to the least cost is optimal.

The proposed method also applies to passenger transportation problems, although in this case it is somewhat simpler since the partitions are determined by the passengers and not the analyst. A good example of this type of application is Wirasinghe et.al. (1977). This reference examines an idealized situation in which passengers traveling to a city from its outlying suburbs have the option to travel either directly by bus (no transshipments) or indirectly by a faster transit system whose stations can be accessed by means of feeder bus lines. Passengers are assumed to use the fastest travel option so that proximity to the transit stations and distance from the city are the main determinants of their choices. Therefore, the resulting partitions are purely geographical. The reference is noteworthy because it appears to be the first application of the CA approach to this type of problem.

Items with different densities: The decomposition approach can also be used when customers differ in other ways. We have assumed so far that an "item" is a given volume of a commodity and that a vehicle can hold a fixed number of items, v_{max} . Although this is a fair description for most freight, for some commodities a vehicle will exceed the roadway axle-weight limitations before it is filled. To side-step this problem we can define an item as a unit of weight when the commodity being handled is denser than the ideal density for the vehicle (the ideal density is the ratio of the vehicle's weight and volume capacities). All the discussion, theory and methods presented up to this point also hold for dense commodities without any modifications.

This of course assumes that all the destinations request items of similar density, or at least denser than ideal. If some customers are "light" (requesting items lighter than ideal) and others are "heavy" (denser than ideal) it may be advantageous to use an asymmetric treatment to exploit the differences. This situation is more likely to arise for collection problems from many suppliers than for distribution problems; e.g. for the collection of the many different parts needed at an automobile assembly plant such as foam for seats, nuts and bolts.

If a single origin produces many different commodities for a single destination, it is not difficult to see that the number of vehicle loads needed to carry the amounts produced in a given time is minimized if one of the fol-

---

[5] This argument is given in more detail in Daganzo (1985a) as a basis for the dynamic program proposed in that reference.

lowing two conditions is satisfied (Daganzo and Hall, 1985): (i) either all vehicles reach their weight capacity, except possibly the last one which may be partially filled, or (ii) all the vehicles reach their volume capacity. Note that one of the conditions is sure to be satisfied if all the loads are as large as possible while roughly containing the same mixture of items. This should be clear since all the loads will then be either below or above the ideal density. As a corollary of this observation we note that if a single destination is fed without transshipments from many small suppliers producing different items, then the symmetric collection strategies described in Chapter 4 also minimize the number of vehicle tours. This happens because if both "heavy" and "light" suppliers are uniformly scattered over the area, then the collecting vehicles will automatically tend to pick up item mixtures with approximately the same density. Without transshipments, thus, there seems to be no incentive to discriminate across customers. An exception occurs if light and heavy suppliers tend to form separate clusters, as illustrated in the above reference; in that case it may be advantageous to increase the length of some tours to enhance their load composition, thereby reducing their number. This is in general a complicated problem, whose accurate solution depends on details such as the relative proximity of light and heavy supplier clusters.

Another exception occurs if transshipments are allowed. A case of particular interest occurs if the vehicles are only allowed to make one stop, but collection can take place with a transshipment. Suppliers with the lightest and densest commodities have the most to gain from sending their shipments through the terminal since, combined with complementary commodities at the terminal, they can be carried to the destination in ideal density loads requiring fewer vehicle-miles. An asymmetric treatment of customers would then be in order. Problem 5.9, based on Daganzo (1988), encourages the reader to develop an optimal asymmetric shipping strategy where the rent for space and the items are so cheap that holding costs can be ignored; only transportation and handling costs need to be considered. The solution is obtained by decomposition, conditional on the number of ideal density truckloads sent through the terminal. As part of the solution, the reader needs to determine which suppliers – and how much of their production – should be shipped through the terminal to obtain the conditioning flow through the terminal.

5.5 Multiple Transshipments

Multiple transshipments are unlikely to be advisable for most physical distribution applications, because each additional transshipment generates additional handling costs and the vehicle economies (v_{max} vs. v'_{max}) can be achieved with just one transshipment. In any case, systems that allow multiple transshipments can be designed, using the one-transshipment results as a building block. This section, based on Daganzo and Newell (1986), presents a simple recursive technique to this effect, and illustrates it with an example. The technique uses the function $z_1(\lambda,r,\delta)$ – depicted in Fig. 5.5 and obtained from the minimum of (5.5), (5.9) or similar expressions – to construct a function $z_2(\lambda,r,\delta)$ representing the minimum cost per item with at most two transshipments.

Figure 5.3 depicts a level-1 terminal and its influence area, whose size is now denoted $I_1(\mathbf{x})$. Recall that all the customers in a level-1 area are served from the level-1 terminal with at most 1 transshipment, not including the one at the level-1 terminal, and that the level-1 terminals themselves are served without transshipments from the depot. This structural organization makes it easy to express, conditional on I_1 , the inbound, outbound and terminal costs for a level-1 terminal in a form similar to Eq. (5.5); the logistic cost function is now:

$$\begin{pmatrix} cost \\ per \\ item \end{pmatrix} = \underset{terminal\ cost}{\left(\alpha_5 + \alpha_6/I_1\right)} + \underset{inbound\ cost}{z^i\left(\lambda,r,I_1\right)} + \underset{outbound\ cost}{Z^1\left(\lambda,r,I_1\right)}. \qquad (5.11)$$

The terminal and inbound costs assume the same functional form as in (5.5), since the cost of delivering and passing through the level-1 terminals does not depend on how the items are treated once they leave them. The outbound cost, however, is different. It has been superscripted by "1" since Z^1 should now represent the average of $z_1(\lambda,r,\delta)$ instead of the (larger) $z_0(\lambda,r,\delta)$ – see Fig. 5.5.

As in Eq. (5.4) we may want to approximate the average cost by the cost of the average:

$$Z^1\left(\lambda,\delta,I\right) \cong z_1\left(\lambda, 0.38I^{1/2}, \delta\right), \qquad (5.12a)$$

but the accuracy of this approximation will now have deteriorated because z_1 is more highly non-linear as a function of r than z_0 – see Fig. 5.5. One may instead opt for using the exact definition:

$$Z^1(\lambda, \delta, I) = E_r[z_1(\lambda, \delta, r)].$$ (5.12b)

Either one of these expressions can be used to find the minimum of Eq. (5.11) with respect to I_1. The result should be a function of λ, r, and δ, $z_2^*(\lambda, r, \delta)$ which, as a function of r, should start higher and be flatter than either z_0 or z_1^*. As a result, we may find a second critical distance beyond which two transshipments are needed ($z_2^* < z_1$). For most practical problems, though, this distance is likely to be large compared with the distance between the depot and the farthest reaches of **R**.

It is theoretically possible, but practically unnecessary, to iterate this procedure to obtain the optimal size of higher level influence areas. The technique can also be applied if shipments are to be synchronized at the level-1 terminals,[6] and also if constraints require a more extensive list of conditioning variables for the decomposition principle to apply.

In order to design the system one would carve out the service region into influence areas approximating the ideal size $I_1(\mathbf{x})$ as was illustrated in Fig. 5.6. Of course, this only needs to be done for the portion of **R** lying beyond the second critical distance. The headways at the level-1 terminals, a byproduct of the optimization, can be used to construct the level-1 feeder routes and schedules. Within each level-1 influence area, the system can be designed as in Secs. 5.3 and 5.4. An example illustrates the procedure.

5.5.1 Example

The example that led to Eqs. (5.7) is continued here. However, to simplify the notation we will give some arbitrary values to the constants that appeared: $v_{max} = b = b' = a/\lambda = a'/\lambda = 1$, and will then eliminate these variables from the notation. We assume that the demand and customer density do not depend on location or time, and use the case with $v'_{max} = \infty$.

We already know that:

$z_0 = 1 + 2.7\,r,$ from Eq. (5.6a)

$Z^0 = 1 + I_0^{1/2},$ from Eq. (5.6b)

$z^i(r, I_0) = 2(r/I_0)^{1/2},$ from Eq. (5.6e)

$I_0^* = 2\,r^{1/2},$ from Eq. (5.6f) and

---

[6] In this case one would minimize (5.9) *holding H^i constant*, and this variable would appear in the expression for Z^1. The counterpart of (5.11) would then include the inbound and outbound headways as decision variables, in addition to I_1.

$$z_I^* = 1 + 2.8 \, r^{1/4}, \qquad \qquad \text{from Eq. (5.7b).}$$

Thus,

$$z_I(r) = 1 + \min \{2.7r; 2.8r^{1/4}\}, \qquad \qquad (5.13)$$

which is plotted on Fig. 5.8. Note that when $r \geq 1.05$, transshipments become necessary. To calculate $Z^1(I_1)$, one should take the expectation of (5.13) for the r values that arise in an influence area of size I_1: $r \in [0, (I_1/\pi)^{1/2}]$.

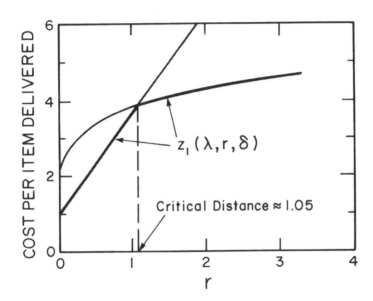

Fig. 5.8   Example: cost of serving a point with and without a transshipment. (Adapted from: Daganzo and Newell, 1986)

For small influence areas ($I_1 \leq 0.29^{-1}$), $z_1(r) = 1 + 2.7r$ and

$$Z^1(I_1) = 1 + I_1^{1/2}, \qquad \text{if } I_1 \leq 0.29^{-1}. \qquad (5.14a)$$

For $I_1 \geq 0.29^{-1}$, we find:

$$Z^1(I_1) = 1 + 2.7 \int_0^{1.05} \frac{r \, 2\pi r}{(I_1/\pi)^{1/2}} \, dr + 2.8 \int_{1.05}^{(I_1/\pi)^{1/2}} \frac{r^{1/4} \, 2\pi r}{(I_1/\pi)^{1/2}} \, dr$$

$$= 1 + 2.17 \, (I_1^{1/8} - I_1^{-1}) \quad \text{if } I_1 \geq 0.29^{-1}. \qquad (5.14b)$$

If (5.14a) applies, the level-1 influence area is not large enough to require another transshipment. We expect (5.14b) to apply for the optimal I_1^*. Equation (5.11) now becomes (remember that we assumed $\alpha_5 = \alpha_6 = 0$) :

$$2(r/I_1)^{1/2} + 1 + I_1^{1/2} \qquad\qquad \text{if } I_1 < 0.29^{-1} \qquad\qquad (5.15a)$$

$$2(r/I_1)^{1/2} + 1 + 2.17(I_1^{1/8} - I_1^{-1}) \qquad \text{if } I_1 \geq 0.29^{-1}. \qquad\qquad (5.15b)$$

This expression should now be minimized for all values of r.[7] Figure 5.9 plots the reciprocal of I_1^* as a function of r. Figure 5.10 plots the minimum cost as a function of r as well.

When r reaches 3.75, the cost, z_2^* , equals z_1 . For larger values, two terminal shipping is best.

Fig. 5.11 depicts a possible configuration of influence areas for a square of side 7 that attempts to be true to the density of terminals shown in Fig. 5.10. Unfortunately, the size of the influence areas forces them to include points that would be better served with larger or smaller influence areas. For example, the level-1 influence zones have an area of approximately 20 units, but they include points that optimally would require $I_1 = 13$ to $I_1 = 42$, plus a few corners with even more stringent requirements. Inspection of Eq. (5.15), however, reveals that variations from the optimal I_1 by a factor of 2 only increase the objective function by about 1 percent. (This robustness is even more pronounced than that observed for level-0 influence areas because the exponents of the objective function are now closer to zero). Thus, the departures from optimality observed in Fig. 5.11 should not matter much.

---

[7] For this particular problem the task is easy. One can find for every I_1^*, the value of r that makes it optimal – and one can be plotted against the other.

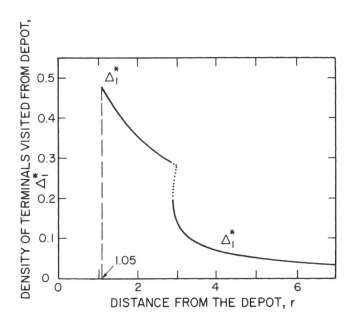

Fig. 5.9   Example: solution to Eq. (5.15). (Adapted from: Daganzo and Newell, 1986)

The exact location of the boundaries and terminals can be fine tuned if desired, but since they are fairly round and centered, respectively, the configuration shown should be nearly optimal. In fact, even the precise location of the boundary between 2 and 1 transshipment service areas is not particularly crucial. See Fig. 5.10. The following section describes an automatic way to fine-tune, or even develop a design.

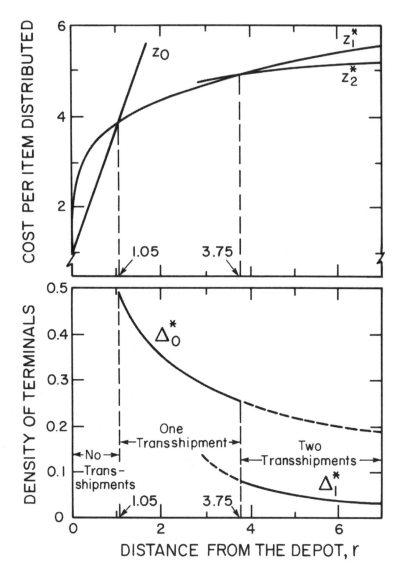

Fig. 5.10   Example: optimal density of terminals as a function of distance. Regions where zero, one, or two transshipments are best. (Adpated from: Daganzo and Newell, 1986)

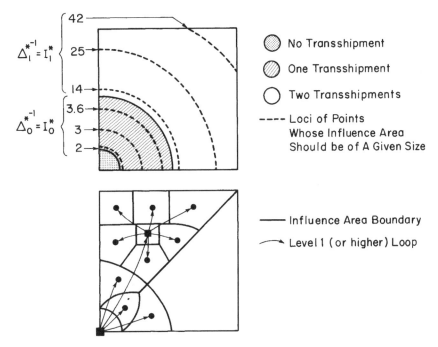

Fig. 5.11   Example: partition of service zone into influence areas. (Source: Daganzo and Newell, 1986)

5.6 Automatic Discretization

Before starting, we should mention that the design problem has also been treated in the scientific literature as a pure optimization exercise - without resorting to the CA approach. In the applied mathematics literature the problem is called the "optimal resource allocation problem;" see Okabe et al. (1992) and Du et al. (1999). Pertinent works seek cost-minimizing locations for point-like service facilities in a space continuum, among a continuum of customers. Unfortunately, these optimization problems turn out to be "easy" only when cost is defined as a simple function of a distance norm. This cost structure, e.g., with the translational symmetry implied by a norm, is unrealistic for typical logistics problems where costs are complicated and almost invariably location-dependent.

More realistic cost scenarios can be analyzed by considering discrete versions of the problem with only a finite number of locations. An extensive operations research literature explores this line of inquiry; see e.g., Daskin (1995) and Drezner and Hamacher (2002). Problems of this type are usually solved with mixed-integer programming techniques, where the terminal locations and customer allocations are decision variables. But unfortunately, existing programming methods can only deal effectively with small problems if they have complicated cost structures.

Our manual method overcomes these drawbacks. It succeeds, e.g., as in the example of Sec. 5.5.1, because it decomposes the problem in two manageable parts. We first look for a continuous target $I^*(x)$ without paying attention to the discrete locations, and then delegate the difficult but non-crucial task of finding the specific locations to the human mind. As explained in Sec. 5.3.4, the human designer is simply asked to partition the service region into "round" influence areas $\{I_i\}$ of a size consistent with the CA target $I^*(x)$, and a set of centrally-located terminals $\{x_i\}$.[8] The remainder of this section, based on Ouyang and Daganzo (2004), shows that this second step can also be performed automatically, even for large problems.

Because roundness is important, we first look for a set of non-overlapping circular disks contained within the service region, of individual sizes as close the ideal $I^*(x)$ as possible. The number of disks is given by the CA procedure: $N = \int I^*(x)^{-1}dx$. More specifically, if we characterize the disks by their centers x_i and their radii r_i (for i = 1, 2, ... N), we look for a set of (x_i, r_i) that satisfy: $I^*(x_i) \approx k\pi r_i^2$ for i = 1, ... N, for a value of k as close to 1 as possible. Once this is done, we generate influence areas by allocating each point in the service region to the nearest x_i. This is the right thing to do because it guarantees that the influence areas so generated contain one disk a piece. Therefore, they must be "round" - assuming that a solution with $k \approx 1$ has been found.

To find a set of disks, we assign some initial values to the (x_i, r_i) and model the disks as if they were physical particles that (i) are repelled when they overlap either with each other or with the boundary, and (ii) change radius as they move over the service region with the recipe: $r_i \approx [I^*(x_i)/k\pi]^{1/2}$. If k is sufficiently large, a discrete-time simulation of this system quickly leads to an equilibrium where all forces vanish and there is no overlap.[9] The simulation is then repeated with a smaller k. A step-wise gradual reduction in k is continued until an equilibrium cannot be found. This will happen before k =1, since circles do not partition Euclidean space. The procedure is then terminated.

---

[8] In this section, the subscript of the influence area variable "I" no longer refers to the influence area level.

[9] This assumes that the service region is "simply connected", in the sense that a disk of proper size can always be slid between any two points in the service region without touching the boundary. No generality is lost by this assumption, because complex areas (e.g., Japan) can usually be partitioned into simply connected components to which the model can be applied separately.

It is reported in Ouyang and Daganzo (2004) that this procedure can quickly find good designs to problems of practical size. Figure 5.12 shows how the method converges in a case where the best design is known. The region is poly-hexagonal with N= 7, and the target area size $I^*(\mathbf{x})$ is independent of location. The best design is shown in Fig. 5.12(a), and the algorithm's results in parts (b)-(d).

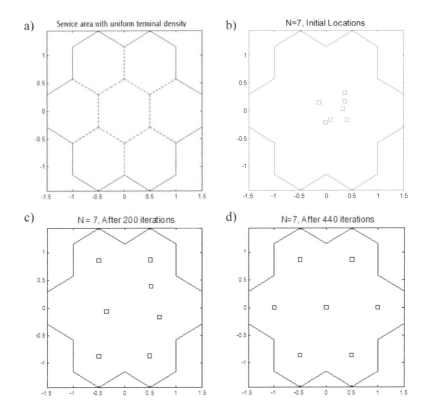

Fig. 5.12   Verification of convergence: (a) ideal solution, (b) initial locations, (c) location after 200 iterations, (d) equilibrium after 440 iterations (Source: Ouyang and Daganzo, 2004)

The algorithm has also been applied to the example in Sec. 5.3.5 using (5.6f) as the target function with a = b = a' = b' = v_{max} = 1; i.e.: $I^*(\mathbf{x})$ = $2[r(\mathbf{x})/\lambda(\mathbf{x})]^{1/2}$. (Recall that $r(\mathbf{x})$ was the Euclidean distance to the depot, and $\lambda(\mathbf{x})$ the demand density.) Two cases were considered: (a) uniform demand, where $\lambda = 1$ and $I^*(\mathbf{x}) = 2r(\mathbf{x})^{1/2}$; and (b) declining demand, where $\lambda(\mathbf{x}) = r(\mathbf{x})^{-1/2}$ and $I^*(\mathbf{x}) = 2r(\mathbf{x})^{3/4}$. Figures 5.13 and 5.14 show the respective results for four square regions of sides L = 5, 7, 10 and 25. All the examples were solved in less than 30 minutes on a 1.7 GHz PC.

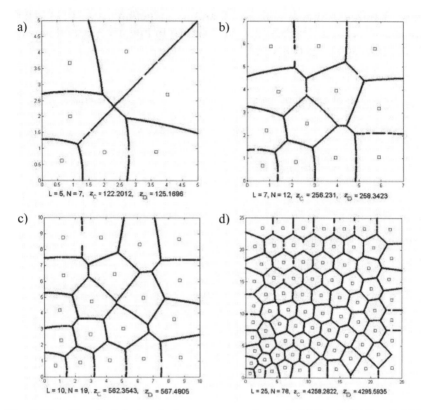

Fig. 5.13   Terminal design for homogeneous demand: (a) L – 5, (b) L – 7, (c) L – 10, (d) L – 25. (Source: Ouyang and Daganzo, 2004)

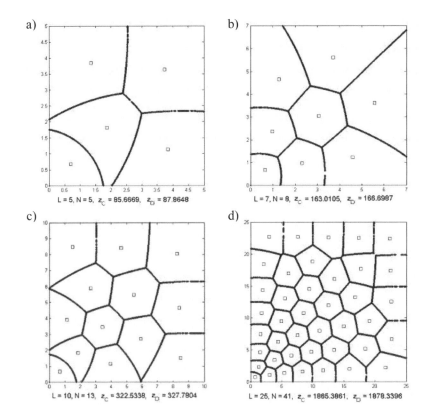

Fig. 5.14   Terminal designs for homogeneous customer demand: (a) L – 5, (b) L
– 7, (c) L – 10, (d) L – 25. (Source: Ouyang and Daganzo, 2004).

In the uniform demand case the difference between the CA cost prediction
for the variable costs – the integral of the last term of (5.7b) over the service region – and the variable costs arising from the design is quite small:
2.4% for L = 5, 0.8% for L = 7, 0.9% for L = 10, and 0.9% for L = 25. In
the variable demand case the cost differences are 2.6%, 2.3%, 1.6%, and
0.7% respectively. All these differences are exaggerations because they ignore fixed costs, such as the first term of (5.7b), which are large and can be
predicted without error by the CA method. In all cases, the CA prediction
was lower than the actual cost. Ouyang and Daganzo (2004) further argue
that this is not a coincidence, and that the CA predictions for our examples
should be lower bounds to the optimum solution. Thus, the percentage differences we observed can be interpreted as optimality gaps.

Note that in both scenarios, and in agreement with theory, the accuracy
of the CA formulae and the efficiency of the proposed design method im-

proves with problem size considerably. This is fortuitous. It means both, that the CA formulae describe well the optimum costs of large complex problems, and that the CA discretization algorithm can complement conventional optimization methods when they would have the most difficulty.

Although the discretization procedure was illustrated with Euclidean metrics, it can also be applied to other metrics by deforming the disks during the simulations, and using true distances in the tessellation step. For example, designs for L_1 metrics should use square "disks" with the same repulsive forces as before, and the L_1 distance formula. An example is shown in Figure 5.15.

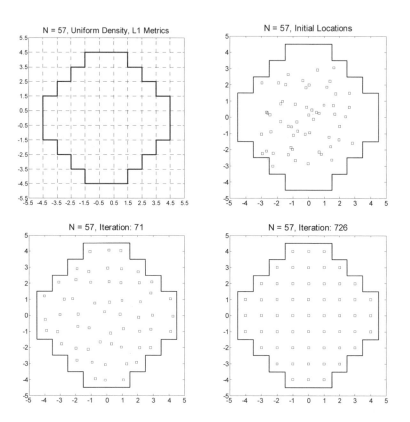

Fig. 5.15   Performance with an ideal solution.(Source: Ouyang and Daganzo, 2004).

Suggested Exercises

5.1 Consider a square influence area centered at the origin in the cartesian plane, with side 2. Travel on the plane can only take place in direc^-tions parallel to the coordinate axes. Both the depot and a terminal are placed on the abscissa axis. The depot is located at x= −10. The terminal is located at $x = x_T$.

Vehicles from the depot only visit the terminal. Vehicles from the terminal visit K randomly scattered points on the square once a day. Each day, one vehicle arrives at the terminal from the depot and K trips leave the terminal to a different set of points.

Determine an expression for the expected total number of vehicle-miles traveled per day as a function of $K \geq 1$ and $-1 \leq x_T \leq 0$. Plot the relative difference between the minimum of this expression and its value for $x_T = 0$ as a function of K .

Demonstrate that for a Euclidean metric the relative difference is similar. (Suggestion: the problem can be solved analytically, but this is tedious. Instead, the reader may want to generate K random locations and calculate the differences in total distance with pencil and paper; a computer spreadsheet greatly simplifies the task).

5.2 Consider the VRP problem as described in Sec. 4.2 with distances given by a Euclidean metric and points homogeneously distributed on a lattice. Assuming that the region contains many more vehicle tours than points per tour (i.e. tours can be elongated as desired), show that the "k" coefficient is 0.5 for a square lattice, slightly larger ($12^{-1/4}$) for an equilateral triangle lattice, and slightly smaller ($27^{-1/4}$) for a hexagonal lattice. (See the application in Daganzo, 1984b, for a hint). (In all cases it is smaller than the coefficient for randomly distributed points.)
Repeat the analysis for the TSP (the result then applies to the VRP when there are few vehicle tours). Is the "k'" coefficient larger or smaller now?

5.3 Repeat the example of Section 5.3.5 assuming that schedules can be synchronized, and that $c'_h = c_h$ and $c'_d = c_d$. What would happen if $v'_{max} = v_{max}$? Discuss from the solution how the benefit of coordination depends on the ratio v'_{max}/v_{max} , the item value, the cost of transportation, etc...

5.4 Consider again the example of Sec. 5.3.5, with the numerical data used in Sec. 5.5: $v'_{max} = \infty$; $v_{max} = b = b' = a'/\lambda = a/\lambda = 1$. Assume that (also as in Sec. 5.5), the depot is located on one of the corners of a square of side 7. Generate within that square two random sets of available terminal locations, with 5 and 25 points each. Compare the cost and structure of the solutions you would recommend in both cases, both to each other, and to the cost predicted by the CA approach without restrictions to location.

Redesign the system if one half of the square contains 2 or three terminals (e.g. corresponding to the locations of the set with 5 locations), and the other half about 12 or 13 terminals (corresponding to locations from the other set).

5.5 Consider again the elevator system design of Problem 4.8. Describe and apply a procedure to find the optimal location for a skylobby, the number of elevators allocated to shuttle service, and the banking strategy for the building sections above and below the skylobby. (Elevators provide shuttle service between lobbies, so that a passenger traveling to one of the top floors will take a non-stop elevator to the skylobby and then will transfer to a "local" elevator to the final floor; see Fruin, 1971, and Mitric, 1972, for descriptions).

5.6 The demand of a hypothetical customer follows a normal stochastic process with independent increments. The demand rate is D' and the index of dispersion is γ (i.e. the variance of the number of items consumed in time T is: $\gamma D'T(items^2)$). Possible delivery times are at time instants, $t = 0, H, 2H, 3H, ...$ If an order is placed at time t_o, it arrives at the delivery instant immediately after: $t_o + T_\ell$, where T_ℓ is an order response time (or lead time).

Suppose that the customer places an order when the inventory on hand plus the amount on order jointly equal a trigger value, v_o (more complicated strategies can be devised depending on *when* the trigger point is reached, but they would be harder to implement). The size of the order is a constant v. (For $H = 0$, this strategy is analogous to the one presented in Chapter 2 in connection with Fig. 2.9.)

Assuming that $v \gg D'H$ (i.e. we order enough to last several headways), and that v_o is chosen to avoid stock-outs, prove that the average (and maximum) accumulation at the destination decompose as suggested in the text: a headway delay term equaling $\frac{1}{2}H$ (and H), a

load make up term proportional to v and independent of H, and a safety stock term independent of v.
(**Hint:** show that if v_o is chosen to equal $D'(H + T_\ell)$ + (CONSTANT)$(\gamma D'(H + T_\ell))^{1/2}$, stockouts are very unlikely for (CONSTANT) > 2. Then prove the statement).

Show as well that the proportion of delivery points where there is a stop is: $D'H/v$, and that given a reorder cost, c_p, the optimal proportion is a simple function of c_p , c_i , D' (constants) and H.

5.7 A factory produces items for a region containing many small customers. Customer n consumes items at an average (stationary) rate, D'_n , but the actual number of items demanded during time intervals of length T days change over time. The actual number in any interval is viewed as a random variable with mean $D'_n T$ (items) and variance $\gamma D'_n T$ (items^2), where γ (items) is a constant index of dispersion which is the same for all the customers. The amounts consumed in any collection of non-overlapping intervals are mutually independent variables, independent of the amounts consumed by other customers. For intervals such that $D'_n T >> (\gamma D'_n T)^{1/2}$, the variables are approximately normal.

Two transportation modes are available. Mode 1 is fast but expensive; items can be considered to arrive at their destination a short time after they are ordered, but there are no economies for transportation in bulk – an item using this mode from point A to point B pays a fixed rate, c_1 ($/item-mile), based on distance AB. Mode 2 is slow but cheap.

Currently only mode 2 is used to deliver items to the customers, but due to its slowness the customers must carry too large a safety stock (see Fig. 2.9). We seek to evaluate the merits of introducing a system of warehouses, similar to the level-0 terminals of this chapter, whose function would be to consolidate the demand fluctuations of all the customers in each influence area. Items would be delivered to the warehouses by mode 2 and from the warehouses with mode 1. This would effectively move all the safety stock from the customers to the warehouses – reducing its total size – but would increase transportation cost since mode 1 is now used.

Assuming that items are delivered non-stop from the origin to each terminal in a full truck carrying v_{max} items, find the size I, of a warehouse's influence area which minimizes the total logistic cost. You may use c_i for the waiting cost per item-day and assume that in any subregion P of R of a size comparable with likely sizes of an influence area, the number of customers with demand D'_n is $\delta_n|P|$ and all the customers are uniformly distributed.

(**Hints:** If the average distance from the depot to a warehouse is independent of the number of warehouses – a reasonable assumption – then the cost of mode 2 transportation and handling is independent of the number and location of terminals. Explain why. You will need to determine the characteristics of the demand at a warehouse as a function of I, which will yield the inventory cost at the terminal from a construction such as Fig. 2.9. You will also need to estimate the number of item-miles traveled in the influence area.)

5.8 Repeat the example of Sec. 5.3.5. for the case $v'_{max} = \infty$, assuming that one half of the demand comes from a few large customers, "a" = 0.1 , (remember a = $c_h\delta$) and the other half from a larger number of smaller customers with "a" = 10 . Also assuming that $\alpha_6 = 0$, explore the nature of the solution for various values of α_5 – the transshipment penalty per item. Consider discriminatory customer treatment strategies.

5.9 Items from many origins, i , have to transported to a common destination, D_i miles away. A single organization (presumably located at the destination) is responsible for coordinating their transportation, and for the resulting freight charges. Although all of the flow, q_i , from each origin can be shipped directly to the destination, it is also possible to ship a portion of the flow, f_i , (possibly all of it) to a terminal, d_i miles away, where it can be dispatched to the final destination. The distance between the terminal and the destination is D miles. No peddling or collecting is allowed.

Each origin makes items of a unique type, i . Only one vehicle type is available and items are so small that a vehicle can be filled with enough quantity of any item type mixture. Some of the flow from origins producing light items (items that cube out a vehicle before reaching its weight capacity) is consolidated at the terminal with some of the flow from "heavy" origins. Management has determined that the total light-item flow sent to the terminal should precisely combine with the heavy-item flow through the terminal, so that only loads of perfect density leave the terminal.

Let w_i denote the fraction of truck *weight* capacity taken up by one item of type i, and v_i the corresponding fraction of *volume* capacity. The difference $a_i = w_i - v_i$, small compared with 1, will be positive for "heavy" items and negative for "light" items ; it will be zero for items of ideal density. Show then that management's decision is equivalent to the constraint:

$$\sum_i q_i a_i f_i = 0 .$$

Show, as well, that (if this constraint is satisfied) the vehicle-miles saved by using a given set of f_i's ($0 \le f_i \le 1$) can be expressed as:

$$\sum_i q_i b_i f_i ,$$

where b_i is the following (possibly negative) constant:

$$(D_i - d_i)\max(v_i, w_i) - Dw_i .$$

Describe a simple (greedy) algorithm to solve the savings maximiza-tion linear program, and apply it to the data of Table 5.1 ($D = 400$), taken from Daganzo (1988a).

Finally, prove that the "ideal density" strategy adopted by manage-ment is optimal.

Table 5.1

i	$q_i/10$	d_i	D_i	γ_i/γ	$10a_i$	b_i
1	1	10	393	0.23	-0.77	-291
2	1	22	398	0.71	-0.29	-92
3	1	24	399	0.64	-0.36	-119
4	1	42	436	0.32	-0.68	266
5	1	38	392	1.82	0.45	46
6	1	78	333	0.42	-0.58	-87
7	1	100	446	1.32	0.24	54
8	1	96	481	0.31	-0.69	-261
9	1	89	336	1.46	0.32	153
10	1	85	378	1.06	0.06	107
11	1	29	411	1.96	0.49	18
12	1	63	389	1.16	0.14	74
13	1	9	408	1.26	0.21	1
14	1	10	402	1.94	0.48	8
15	1	7	407	1.64	0.39	0
16	1	51	362	1.24	0.19	89
17	1	2	399	1.24	0.19	3
18	1	1	400	0.87	-0.13	-51

Glossary of Symbols

α_1-α_4:	See chapter 4 glossary,
α_5:	Fixed cost per item handled at the terminal ($/item),
α_6:	Eq.(5.3) constant, capturing terminal scale economies,
c_d:	Cost per vehicle-"mile",
c_f^o:	Terminal handling cost constant ($/time),
c_f^t:	Terminal handling cost per item ($/item),
c_h:	Holding cost per item-day,
c_i:	Inventory cost per item-day,
c_p:	Price charged for a customer delivery,
c_r^t:	Rent per item-day at the terminal,
c_r^o:	Minimum rent per terminal,
D':	Item flow through a point (customer or terminal),
D'_n:	Item flow for customer n,
δ:	Spatial customer density (customers/area),
δ_e:	Effective spatial customer density (stops/area),
f_o:	Fraction of customers treated differently,
γ:	[Variance/Mean] of items demanded (index of dispersion),
H:	Headway,
H*:	Optimum headway,
H^i:	Inbound headway,
H^o:	Outbound headway,
H^o_j:	Outound headway to sector j,
H^t:	Fixed terminal transfer time,
I:	Influence area size (level-0),
I_j:	Level-j influence area size,
I*:	Ideal influence area size,
I_{min}:	Minimum allowable influence area size,
I_{max}:	Maximum allowable influence area size,
k:	Dimensionless constant of the VRP,
λ:	Demand density rate (items/time-area),
n_s:	Number of stops per tour,
n_s*:	Optimal number of stops per tour,
P:	Subregion of **R**,
r:	Distance from the depot,
R :	Service region,
t:	Time,
t^{max}:	Time allowed for distribution,
t_1^{max}, t_2^{max}:	Time allowed for distribution: (1) inbound, (2) outbound,
T_ℓ:	Order response time,
v:	Delivery lot size,
v^o:	Inventory trigger point,

v^*:   Optimal customer delivery lot size (Sec. 5.4.3),

v_{max} , v'_{max}: Vehicle capacities (items),

$\mathbf{x}$:   Spatial coordinates of a point,

$z^i(\lambda,r,I)$:   Inbound cost per item,

z_m^i: Inbound motion cost per item,

z_m^{i*}:   Inbound motion cost per item, for an optimum headway,

z_m^o: Outbound motion cost per item,

z_m^{o*}:   Outbound motion cost per item, for an optimum headway,

$z^o_{m,j}$:   Outbound motion cost per item to sector j,

$z_j(\lambda,r,\delta)$:   Cost per item distributed with j, or less, transshipments,

$z_j^*(\lambda,r,\delta)$: Cost per item distributed with j transshipments,

$Z^0(\lambda,\delta,I)$:   Average outbound cost per item in an influence area of size I ; no transshipments,

$Z^j(\lambda,\delta,I_j)$: Average outbound cost per item in an influence area of size I_j ; j transshipments,

$Z^1_T(\mathbf{P})$:   Cost per unit time in $\mathbf{P}$, with one or less transshipments.

6 Many-To-Many Distribution

Readings for Chapter 6

Of all the chapters in this monograph, this is perhaps the one which is least related to the existing literature. Sections 6.2, 6.4.2 and 6.5.1 - considering respectively distribution with 0, 1, and 2 transshipments - are based on Daganzo (1987c). The results in this reference, however, are not presented literally; a more refined routing scheme, developed in the chapter, is used here. The discussion on detailed asymmetric strategies for distribution with only one terminal, at the end of Sec. 6.3.2, is based on Blumenfeld et.al. (1985a) and Hall (1987).

6.1 Initial Remarks

The first five chapters of this monograph have been devoted to logistics problems involving the movement of freight and people from one origin to any number of destinations – or else to the reverse problem of collecting freight and people from any number of origins for a single destination.

We now turn our attention to problems involving any number of origins *and* destinations. In practice, many-to-many problems arise in connection with public carriers such as: airlines, the postal service, less-than-truckload carriers, railroads, etc. Unlike for private carrier operations, where most of the logistic costs are borne by the firm, some costs are now borne by the carrier and some by the shipper. This dichotomy, introduced in Sec. 5.4.3 for the one-to-many problem, will also be recognized here.

The logistic problem will be specified as before, in terms of a geographical distribution of origins and destinations with certain supply and demand rates. It will be assumed that each destination demands a specific number of items from each one of the origins and that these cannot be substituted for one another. That is, we are dealing here with what normally is referred to in the network optimization literature as a *multi-commodity* problem.

Single commodity problems arise if destinations specify a combined demand regardless of point of origin. Examples of these are water and electricity supply problems. These problems allow the analyst to specify which origins ship to which destinations, which greatly reduces the need for travel and transportation.

Single commodity problems are not addressed in this chapter because they can be reduced to special cases of the problems studied in Chapters 4 and 5. If there are more destinations $\{j\}$ than origins $\{i\}$, one can introduce a *single* "super-origin" O_o with production rate equal to the total demand rate at $\{j\}$, and then worry instead about finding the best scheme for serving $\{j\}$ from O_o through a set of "terminals," $\{i\}$, assuming that items can be moved freely from O_o to $\{i\}$. To ensure that each real origin ships the prescribed amounts, the capacities of the fictitious terminals should be set equal to the origin's maximum production rates. If there are more origins than destinations, one would seek the best way of carrying items from the origins to a fictitious super-destination D_o.

This one-to-many interpretation of the single commodity problem indicates that each real origin should serve the destinations in an influence area surrounding it, possibly with a transshipment, that these influence areas partition **R**, and that – conditional on certain variables – the operations in the influence areas should be independent of one another. Of course, the shed boundaries separating influence areas may shift with time if the demand and production rates vary with time. The system can be designed as described in Secs. 5.4 and 5.5.

Unfortunately, multi-commodity problems cannot be reduced in the same manner to a problem with a single origin or destination; they are inherently different and more difficult. This chapter will focus on the aspects of the problem that are better understood, for the most part involving stationary data and solutions, and emphasizing problems for which pipeline inventory is a negligible quantity. Reasonable for most freight transportation problems, this emphasis may not be appropriate for many-stop passenger transportation systems, such as public transit. A thorough treatment of passenger transportation issues is beyond the scope of this monograph.

This introductory section concludes with a brief discussion of a new role played by terminals in many-to-many logistics systems; Sec. 6.2 then examines distribution without terminals and no transshipments. The following two sections examine the organization of systems with only one transshipment per item: Section 6.3 discusses one-terminal systems, and Sec. 6.4 systems with several terminals. The organization of these systems is discussed at three levels: (i) an operational level where the dispatching schedules and terminal locations are given, and we only seek the best way of routing the items and the vehicles; (ii) a tactical level where only the terminal locations (or only the schedules) are given; and (iii) a strategic level where everything is open to change. Section 6.5 concludes the chapter with a discussion of multi-terminal, multi-transshipment systems.

6.1.1 The Break-Bulk Role of Terminals

Section 5.2 illustrated how transshipments allowed items from a single origin to travel long distances in large batches to a terminal, and then in small batches from the terminal to the customers. This allowed route length restrictions and delivery vehicle size limitations to be met, while preserving transportation economies of scale. The same economies occur in reverse: small vehicles can carry items from scattered origins to a terminal, where the small loads can be consolidated into larger ones for transportation to a destination. We can think of terminals in many-to-one and one-to-many systems as *consolidation* points that allow line-haul and local operations to be decoupled. Consolidation is in fact their only role, for if there are no incentives to keep local routes short (e.g. pipeline inventory considerations, delivery vehicle size limitations, or other restrictions) and a filled vehicle carries a well defined number of items, then transshipments can be shown not to reduce costs (Daganzo, 1988).

This statement can be verified analytically from Eq. (5.5) by: (i) substituting either Eq. (4.10) or Eq. (4.21) with $\overline{D}'$ replaced by λ/δ for the function $z_0\,(\lambda,\,r,\,\delta)$ used in Eq. (5.4); and (ii) using (4.10) or (4.21) with I^{-1} instead of δ for $z^i(\lambda,\,r,\,I)$. Simple algebraic manipulations then reveal that the minimum of (5.5) is greater than $z_0(\lambda,\,r,\,\delta)$.

The statement can also be verified intuitively. If the inbound and outbound schedules can be synchronized, it was shown in Sec. 5.4.1 that the inbound and outbound vehicles should arrive and depart from the terminal in perfect synchronization – with every outbound headway a multiple of the inbound headway. The least cost is often obtained with the scheme of Fig. 6.1a, where all the headways are equal and there is no discrimination across customers.[1] In this case, however, the strategy depicted in Fig. 6.1b is also feasible. To see this, recall that delivery vehicles are as large as level-1 vehicles and that there are no route length restrictions. With the same pickup and delivery schedule as the old, the new strategy preserves holding costs and vehicle mileage. This is true because pipeline inventory costs are negligible. Clearly then, the new strategy is more attractive because it saves handling and terminal costs.

Many-to-many multi-commodity problems are different. Under the same conditions, with no restrictions on delivery vehicle size and route length, transshipments can reduce logistics cost very significantly. Suppose that a near optimal solution for a many-to-many problem without transshipments includes two vehicle routes that: (i) visit the same two sets

---

[1] The case with different headways is discussed in Daganzo (1988). Discriminating strategies for different customer types, some being served through the terminal and some not, also lead to the same conclusion – the construction of Fig. 6.1 can be applied to the set of routes that visit terminals.

of origins and neighboring destinations, as depicted on the top half of Fig. 6.2, and (ii) are operated with the same frequency. This arrangement would be reasonable if the destinations visited are similar as we shall soon see in Sec. 6.2 below.

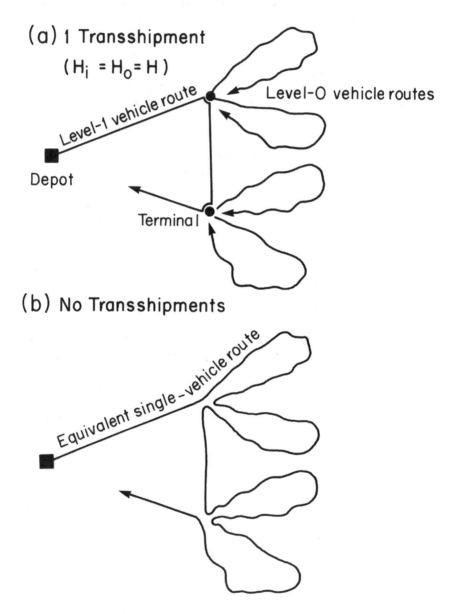

Fig. 6.1   Distribution with and without transshipments

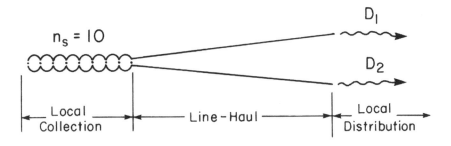

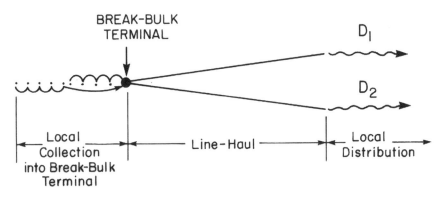

Fig. 6.2   Reduction in collection travel as a result of a break-bulk transshipment. (Source: Daganzo, 1987c)

The bottom half of Fig. 6.2 illustrates how a transshipment can reduce
transportation costs without increasing holding costs at the origins and destinations. Without changing the times of departure, one vehicle could pick
up items for both sets of destinations from some of the stops, and the other
vehicle would do the same for the remaining stops. Both vehicles would
visit just enough stops to carry the same load sizes as in the top part of the
figure. To avoid the need for visiting both sets of destinations with both
vehicles, these would swap appropriate portions of their loads at a terminal
located near the end of their collection runs. It should be clear from the
figure that such a swap would reduce the distance traveled and the number
of stops, without increasing holding costs. It does entail additional fixed,
handling and holding costs at the terminal but if the original number of
collection stops is large the swap could be cost-effective.

Notice that the magnitude of the savings increases with the number of
routes that swap loads at the terminal; e.g., no savings could result with
only one origin, or only one destination. Clearly, this opportunity for sav-

ings is peculiar to many-to-many systems. Because loads must be "broken" before being reconstituted, transshipment points serving this function will be called, consistently with motor carrier jargon, break-bulk terminals (BBTs).

If there are vehicle size or route length limitations, many-to-many systems may include both consolidation terminals (CTs), whose function is to consolidate the small loads carried by local vehicles into larger (long-distance) vehicle loads, and BBTs serving a swapping function for the CTs. This is quite common in existing systems. Motor carrier networks, for example, may include end-of-line terminals (CTs) and break-bulk terminals; railroad networks include industry yards (CTs whose function is accumulating cars from local sidings) and classification yards (BBTs); airline networks include minor airports (CTs) and hubs (BBTs); the postal network is similarly structured, etc. For most of this chapter we will focus on the organization of logistic systems with BBTs only, where only one type of vehicle is used. In the meantime CTs will be viewed as the final origins and destinations. Section 6.5.2 will then show how an integrated system with both CTs and BBTs can be designed.

6.2 Operation Without Transshipments

Let us now examine strategies for serving a collection of scattered origins and destinations without transshipments.

We will restrict our attention for the time being to transportation modes which are not set up to intermingle pick-ups and deliveries within the same route. Not appropriate for passenger transportation, this assumption is reasonable if freight cannot be easily moved within the vehicle. Vehicle tours should then stop at only one origin and multiple destinations – or else the other way around. In this manner the freight does not have to be sorted and restowed every time the vehicle stops for a pick-up or delivery.

Given the spatial densities of origins and destinations $\delta^o(\mathbf{x})$ and $\delta^d(\mathbf{x})$, and an origin-destination flow density $\lambda(\mathbf{x}^o, \mathbf{x}^d)$ denoting the number of items per unit time that need transportation from a region of unit area around $\mathbf{x}^o$ to a region of unit area around $\mathbf{x}^d$ [in this chapter λ has units of items/(time $\times$ distance^4)], we can evaluate the logistics cost per item by comparing 2 strategies:

(i)   peddling with tours from each origin to many destinations,
(ii)   collecting with tours from many origins to each destination.

In the first case, with the transportation, handling and holding cost variables defined as in prior chapters, the cost per item is given by the function "z_0" of Sec. 5.3. The arguments " r, δ, and λ " of this function need to be reinterpreted, though: r now represents the distance between the origin and destinations in a tour (see Fig. 6.3), δ becomes δ^d, and λ must be replaced by λ/δ^o since in earlier chapters λ represented the number of items demanded per unit time and unit area *from one origin*. Similarly, D' should be replaced by $\lambda/(\delta^o\delta^d)$. Thus, for peddling, the cost per item averages: $z_0(\lambda/\delta^o, r, \delta^d)$. In the second case, collecting for one destination, the cost is: $z_0(\lambda/\delta^d, r, \delta^o)$; the solution with least cost should be chosen. If desired one can average this minimum over all possible combinations of $\mathbf{x}^o$ and $\mathbf{x}^d$ to obtain a CA estimate of average cost.

The explicit cost expressions given in Chapters 4 and 5, which would be used to develop z_0, assumed that vehicles return to the depot empty after completing the delivery run. For many to many systems, however, this is unlikely; most vehicles will find loads to carry, if not back to the same origins at least somewhere else. With most vehicles usefully employed at their destinations (load backhauling is discussed in Sec. 6.5.3.), one should discount the cost of the return trips. It is not difficult to see that the cost of open ended trips (as shown in Fig. 6.3) can be captured without changing the form of our logistic cost function, e.g. Eqs. (4.20a) or (4.24a), by using r instead of 2r in the evaluation of the line-haul transportation cost coefficient α_1; i.e. using $\alpha_1 = c_s + c_d r$. We will assume in the following that α_1 has been adjusted to reflect the availability of backhauls.

For a case of peddling cheap items, when pipeline inventory can be ignored, the minimum of Eqs. (4.20) is achieved for $n_s \approx v_{max} (\alpha_4 / \alpha_2)^{1/2}$ (see the derivation of Eq. (4.21)). The minimum cost is then:

$$z_0 = c'_s + \frac{\alpha_1}{v_{max}} + 2\alpha_2 \left[\frac{c_h \delta^d \delta^0}{\lambda \alpha_2} \right]^{1/2}, \qquad (6.1)$$

where we have used $\alpha_4 = c_h \delta^d \delta^o / \lambda$. Recall that α_2 is $c_s + c_d k (\delta^d)^{-1/2}$. Equation (6.1) is identical for collecting except for α_2, which is then $\alpha_2 = c_s + c_d k (\delta^o)^{-1/2}$. Clearly, if $\delta^o > \delta^d$ then z_0 is least for collecting, and the reverse is true if $\delta^o < \delta^d$. This should be intuitive; it implies that the single stop is made at the end of the trip with the largest traffic generator (either an origin or a destination) and that multiple stops are made at the end of the trip where stops are most closely grouped.

With our current definitions for α_2 and α_4, the optimal number of stops leading to (6.1) is:

$$n_s = v_{max}\left(\frac{\alpha_4}{\alpha_2}\right)^{1/2} = \left[\frac{c_h \delta^o \delta^d v_{max}^2}{\lambda\left[c_s + c_d k\left(\delta^d\right)^{-1/2}\right]}\right]^{1/2}. \tag{6.2}$$

The right side of this expression is a dimensionless constant that may depend on x^o and x^d, and we abbreviate it by $K(x^o, x^d)$. It represents the square root of the ratio of two quantities: (i) the average load make-up holding cost per item when every origin-destination pair is served without peddling or collecting by full vehicles, $c_h v_{max}/D' = c_h v_{max}\, \delta^o \delta^d/\lambda$; and (ii) the prorated motion cost per stop of one item in a full vehicle, $(c_s + c_d k(\delta^d)^{-1/2})/v_{max}$. In other words, K^2 can be viewed as the ratio of holding cost to transportation cost for a naive strategy in which one ships in full trucks and allows only one delivery stop; i.e., where only the transportation cost is minimized.

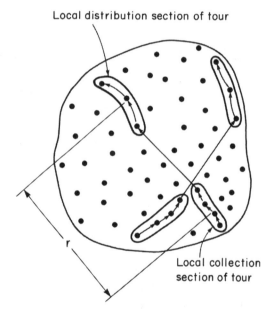

Local distribution section of tour

Local collection
section of tour

r

Fig. 6.3   Many-to-many tours without transshipments. (Source: Daganzo, 1987c)

The quantity "K" can vary by several orders of magnitude depending on the problem at hand and is typically large compared with 1. It can be of order 10^3 (perhaps even larger) when valuable items have to be moved between many small origins and destinations, and it is small when the system consists of few origin-destination pairs with large flows.

The constant "K" allows Eqs. (6.1) and (6.2) to be expressed concisely as follows:[2]

$$n_s = K,$$ (6.3a)

and

$$z_0 = c'_s + \{\alpha_1 + 2\alpha_2 K\}/v_{max},$$ (6.3b)

where α_1 is the line-haul motion cost per trip and α_2 is the motion cost per added stop. If one repeats the analysis we have just done, allowing vehicle tours to make both deliveries and pick-ups, one finds that z_0 increases with the 2/3 power of K and $n_s \approx K^{2/3}$ (Daganzo, 1987c).

Notice that, without transshipments, an unreasonably large number of stops may need to be made. This calls for the introduction of BBTs to shorten vehicle routes. The constant "K" will also appear in the cost expressions for systems with break-bulk transshipments, and will dictate which system configuration is likely to work best. In the following sections, it is assumed that a single vehicle type, with capacity v_{max}, is used. The assumption is relaxed (with the introduction of CT's) in Sec. 6.5.2.

6.3 One Terminal Systems

Figure 6.4 shows how, by linking N_o origins and N_d destinations through a terminal, the number of two-stop routes is reduced from $N_o \times N_d$ to $N_o + N_d$. The reduction is proportionately larger the larger the number of origins and destinations. This helps reduce transportation cost because, with fewer vehicle routes linking origins and destinations, it is possible to carry the same amount of freight with equal service frequency in larger batches with fewer trips. With larger batches, the transportation cost per item-mile can be reduced by a factor: $(N_o + N_d)/(N_o N_d)$. Of course, if a smaller reduction in transportation cost is accepted then the service frequency on all the links can be increased.

This route reduction phenomenon is the basis for the one-terminal strategies explored in this section. Symmetric strategies, where origins and destinations are only differentiated by position within the study region **R**, are studied first. With these strategies nearby origins (and nearby destinations) receive similar service.

---

[2] $z_0 = c'_s - 2\alpha_2/v_{max} + \{\alpha_1 + 2\alpha_2 K\}/v_{max}$ is a more accurate expression when n_s is small.

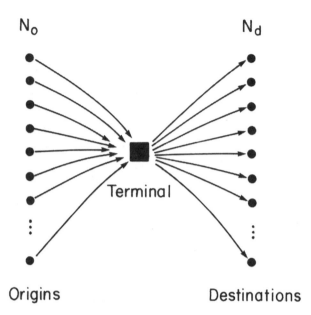

Fig. 6.4   Number of routes with one transshipment

6.3.1 *Symmetric Strategies*

We will explore first the operational level (routing) problem where the terminal location and the service frequency (headways) are given; only the item/vehicle routes are sought. Building on this case, we will then examine the tactical level (scheduling/routing) problem where only the terminal location is given. Finally we shall address the strategic level problem.

The operational problem: With one terminal, the operational level problem is simple. Given a set of inbound and outbound headways – possibly varying across broad subregions of $\mathbf{R}$ – the vehicle routes can be found with the solution of a vehicle routing problem, possibly constrained and including backhauls, as discussed in Secs. 4.2, 4.4 and 4.8.1. This establishes the distance traveled in $\mathbf{R}$ and the total motion cost during each dispatching interval. Since the holding costs are also known, and the number of dispatching instants per day is fixed, the logistics costs per day can be easily estimated.

The tactical problem: At the tactical level we must choose the schedules and decide whether they are to be coordinated at the terminal or not. Without coordination, the terminal costs per item have the form of Eq. (5.3), $\alpha_5 + \alpha_6/|\mathbf{R}|$, which is independent of the tactical variables.

If the inbound and outbound operations are managed as if they were unrelated many-to-one and one-to-many systems then the methodology of Chapter 4 can be used to design each set of operations, and to estimate the resulting average inbound and outbound costs. In the notation of Chapter 4, these costs would be $z_0(\lambda^o, r^o, \delta^o)$, for inbound; and $z_0(\lambda^d, r^d, \delta^d)$ for outbound. Here, r^o (or r^d) represents the distance from an origin (or destination) to the terminal, and λ^o (or λ^d) represents the production (or consumption) rate density in items per unit time per unit area. Clearly, the problem does not require any new treatment.

If schedules can be coordinated, inbound and outbound headways for different subregions of **R** could be chosen from a menu of the form, $\tilde{H} \times 2^p$, where $\tilde{H}$ is an arbitrary time value and p is an integer. This "power-of-two" strategy (Daganzo, 1990) allows the average route to use a headway within 50% of optimal, ensuring at the same time that all the headways are integer multiples or submultiples of each other. As a result, if all the schedules are forced to coincide at one time (say, at time t=0) then every pair of routes will be synchronized as well as possible and the savings from synchronization will be greatest;[3] they will equal the smaller of the two headways on the two vehicle routes used by any given item (see Figs. 4.9 and 4.10). In other words, the third term of (5.9) yields the holding cost, which is a convex increasing function of the headways. This facilitates the design process, as explained below.

Note that the inbound and outbound operations can proceed as described in Chapter 4, as if the terminal was the depot. Therefore, the motion cost can be easily estimated. The notation for motion costs introduced in connection with Eq. (5.9) can be used to summarize these costs. There, we denoted by z^o_m the function relating a terminal's outbound motion cost per item to the demand and customer densities, the size of the influence area I, and the headway H. This function also describes the inbound (or outbound) costs in the neighborhood of **x** for our current situation, if we replace the size of the influence area by $|\mathbf{R}|$, λ by the production (or consumption) rate density in the neighborhood of **x**, λ^o (or λ^d), δ by the density of origins (or destinations), δ^o (or δ^d), and H by either the origin (or destination) headway H^o (or H^d). Because this function must now be used to describe the sum of the inbound and outbound motion costs, the superscript " o " is omitted from the result, which becomes z_m. (In this chapter, the superscript " o " is used to differentiate <u>o</u>rigins from <u>d</u>estinations " d ", whereas it was used to differentiate the outbound from the inbound direction in Chapters 4 and 5).

---

[3] A similar strategy was used in Sec. 5.4.1 for coordinating one inbound headway with many outbound headways; these were multiples of the inbound headway.

If backhauls are used, the dependence of the motion costs on the input/decision variables is qualitatively similar but modifications are needed in the expression of z_m. If the flows in both directions are balanced one can simply use r instead of 2r in the motion cost expressions of Chap. 5, as we suggested earlier. If the flows are unbalanced by more than a factor of 2, a good first approximation is to neglect the cost of overcoming distance for the secondary flow and keep all other costs the same, as explained in Daganzo and Hall (1990). Alternatively, one could use improved formulae such as those proposed in that reference.

In any case, it is now easy to find headways that minimize approximately the sum of z_m and the holding costs. For example, if the "power of two" constraint is ignored, one can find numerically the headways that minimize the (convex) sum of the holding plus motion costs, averaged over the whole region. The headways can then be adjusted without major repercussions, e.g. to the nearest "power-of-2" multiple of the smallest headway. Even more drastically, Daganzo (1990) claims that restricting all the headways to be equal in one of the directions, a helpful simplification, is likely to result in near minimal cost unless there are vast differences among customers. See Problem 6.1.

A case of special interest arises when vehicles are dispatched full because items are cheap, route length is not restricted, etc.... Then, for a given set of headways, z_m is given by the minimum of the second and third (motion cost) terms of Eq. (4.20a) with respect to n_s ; note that the delivery (or collection) lot size v used in the neighborhood of $\mathbf{x}$ is fixed since the headway is given. Because these terms decrease with the number of stops n_s , the minimum is achieved when $n_s v \approx v_{max}$. Therefore, the inbound part of z_m is:

$$(inbound) \quad z_m = \frac{\alpha_1}{v_{max}} + \frac{\alpha_2}{v}.$$

Since $v = (\lambda^o H^o)/\delta^o$ and $\alpha_1 = (c_s + rc_d)$, the inbound motion cost at an origin is:

$$(inbound) \quad z_m = \frac{(c_s + rc_d)}{v_{max}} + \frac{\alpha_2 \delta^o}{(\lambda^o H^o)}.$$

The destination (outbound) cost expression is identical, with the superscript "o" on the last term replaced by " d ".

Note that the above expression increases linearly with r at a rate c_d/v_{max} that is *independent of the headways*. This should not be surprising, since items travel in full vehicles for any headway. Also note that the expression is convex-decreasing in the headways, as stated. Recognizing that the headways can change with position, we can express the total motion cost per day as:

$$\frac{motion\ cost}{per\ day} = \int_R \left[\lambda^o + \lambda^d\right]\frac{\left[c_s + c_d r\right]}{v_{max}}dx + \int_R \alpha_2\left[\frac{\delta^o}{H^o} + \frac{\delta^d}{H^d}\right]dx.$$

Note that the optimal headways will be the result of a trade-off between the second term of this expression and the holding cost per day. Both elements of the trade-off are independent of r, the distance to the depot.[4] That is, *if the terminal is moved, the optimal schedules do not change.* The only portion of the cost that changes is the first integral of the motion cost, which is a weighted average of r across $\mathbf{x}$. This decomposition property will simplify the strategic analysis.

In other words, if vehicles travel full, the optimal cost of the tactical problem depends on the terminal position (through r) as follows:

$$CONSTANT + \int_R \left[\lambda^o + \lambda^d\right]\left[\frac{c_d}{v_{max}}\right]r d\mathbf{x}$$

The same relationship (with a different "CONSTANT") holds true without coordination – see problem 6.2.

The strategic problem: The terminal is optimally located if its distance function $r(\mathbf{x})$ minimizes the total tactical costs.

If vehicles travel full the optimal solution is the minimum of the weighted average expressed by the above integral. This is the well known Weber-point location problem, which can be easily solved (see Losch, 1954, for example). We reiterate that the optimal location is in this case independent of all other operational and tactical details.

If vehicles do not travel nearly full almost always, then the problem does not decompose quite so cleanly. As an approximation, one can calculate the tactical costs for a few candidate locations with different $r(\mathbf{x})$, and make the selection accordingly. This approach should be quite satisfactory.

---

[4] In integral form, the holding cost per day is: $\int_R\int_R c_h \max\{H^o(x'), H^d(x'')\}\lambda(x', x'')dx'\,dx''$

We have argued repeatedly that logistics problems are usually not very sensitive to the specific dispatching times, terminal locations, etc... if those are reasonably close to the optimum. Numerical experimenttation with actual data confirms this for the Weber problem. S. Bhaskaran and R. Kromer (1986) have done extensive sensitivity analyses for locations of General Motors facilities in the continental US, invariably finding that vast regions of the country provide costs within 1% of optimal. Additional evidence in this respect can be found in Campbell (1992 and 1993a).

The same should be true for non-Weber problems. For most large systems, one would expect to find substantial portions of **R** where locating a terminal yields nearly as good a solution as the optimal location. Finding a satisfactory solution, thus, should be easy. Hall (1986) has illustrated the process when time-zones are important and there is a deadline for pick-ups and deliveries. Also arguing that the specific location does not matter much, he shows that the ideal region for locating a terminal is shifted eastward due to the asymmetry introduced by the time-zones. To explore in more detail the concepts of Sec. 6.3.1, the reader is encouraged to solve problem 6.3.

6.3.2 Discriminating Strategies

To this point we have assumed that every origin destination (O-D) pair is served through the terminal. This, however, may be inefficient when the origin is close to the destination and both are far from the depot. Too much circuity is introduced.

Using a conditional decomposition procedure similar to that of Sec. 5.4.4, we examine in this subsection ways of discriminating across O-D pairs. We examine first strategies that only differentiate O-D pairs by their general location within **R,** and then discuss briefly more detailed strategies that account for other characteristics of the origins and destinations.

Decomposition by location: Instead of dealing with O-D pairs individually, they will be treated in groups. To accomplish this, the service region is divided into origin subregions, i, and destination subregions, j, (hopefully not too many); O-D pairs for the same two subregions (i,j) are then treated in the same manner. We use P^i and P^j to denote the surface areas of these subregions. In the following, variables indexed by superscript "i" will refer to origin zones, and variables indexed by "j" to destination zones.

If the inbound and outbound headways at the terminal are known for every i and j, then the operational problem is simple. We need to determine what proportion of the flow from subregion i to subregion j , f^{ij} , should travel direct, and what proportion through the terminal (Figure 6.5). If for direct distribution between subregions i and j it is better to peddle than collect, we will assume that f^{ij} of the origins in subregion i ship to j on a direct, peddling route, and the remaining origins through the terminal. The desired split is then achieved by partitioning the origin subregion into a direct-shipping and an indirect-shipping part. (The same flow split between i and j can be achieved in other ways, but the cost can be shown to be higher). The average cost for items shipped direct, z^{ij}, is given by Eq. (6.1) if vehicles travel full. Note that z^{ij} is a constant independent of all tactical and operational variables. Most notably, it is independent of the f^{ij}. We now show that under certain conditions, the operational problem decomposes by O-D subregion pair.

Conditional on the headways H^i and H^j, the holding cost per item for travel from i to j through the terminal is known and independent of f^{ij}. Recall that it equals $c_h \max\{H^i, H^j\}$ if the schedules are perfectly coordinated.

The motion costs per item ($z_m{}^i$ and $z_m{}^j$) are also as described in section 6.3.1, but with smaller origin and destination flow densities:

$$\lambda^{i} = \sum_{i}\left[\lambda^{ij}\left(1 - f^{ij}\right)P^i\right], \quad \text{for production}$$

and

$$\lambda^{j} = \sum_{j}\left[\lambda^{ij}\left(1 - f^{ij}\right)P^j\right], \quad \text{for consumption}$$

The total terminal-motion cost per day from i to the terminal is thus: $\lambda^i P^i z_m{}^i$. A similar expression holds for the destinations j. Thus, the total terminal-motion cost per unit time is:

$$\sum_{i}\lambda^i P^i\left[z_m^i\right] + \sum_{j}\lambda^j P^j\left[z_m^j\right]$$

If vehicles are dispatched full, we have:

$$z_m^i = \frac{\left(c_s + r^i c_d\right)}{v_{\max}} + \frac{\alpha_2\,\delta^i}{\left(\lambda^i H^i\right)},$$

and the total terminal-motion cost per day is:

$$\sum_{i}\left[\frac{\lambda^i\left(c_s + r^i c_d\right)}{v_{\max}} + \frac{\alpha_2\delta^i}{H^i}\right]P^i + \sum_{j}\left[\frac{\lambda^j\left(c_s + r^j c_d\right)}{v_{\max}} + \frac{\alpha_2\delta^j}{H^j}\right]P^j$$

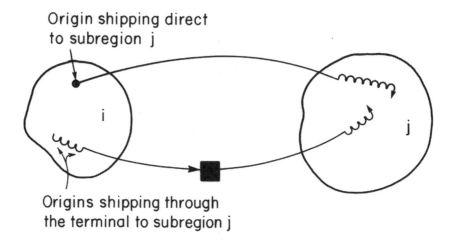

Origin shipping direct
to subregion j

i

j

Origins shipping through
the terminal to subregion j

Fig. 6.5   Routing possibilities, with and without a transshipment

We have used the constants δ^i and δ^j (densities of origins and destinations) in these expressions for the density of collection and delivery stops to/from the terminal. This is reasonable for the origins, in view of the partitioning scheme for splitting the flows, and is also reasonable for the destinations as long as every destination receives some flow through the terminal. Therefore, it is safe to assume that in the expression for the total terminal-motion cost per day only the λ^i and λ^j depend on the splits and that, as explained earlier, the dependence is linear. Note as well that the rate at which cost increases with the splits is independent of the headways, as happened in Section 6.3.1.

If the headways are constant the terminal-motion cost per day is the sum of a constant plus an amount $(c_s + rc_d)/v_{max}$ for every item collected and every item delivered r miles away from the terminal. The contribution of O-D subregion pair (i,j) toward this quantity is:

$$\frac{\lambda^{ij} P^i P^j \left(1 - f^{ij}\right)\left(2c_s + \left[r^i + r^j\right]c_d\right)}{v_{max}};$$

its contribution toward the daily holding-terminal costs is:

$$\lambda^{ij} P^i P^j \left(1 - f^{ij}\right)\left[c_h \max\left\{H^i, H^j\right\}\right];$$

and its contribution toward total daily direct-shipping costs is:

$$\lambda^{ij} P^i P^j f^{ij} z^{ij}.$$

The sum of these three expressions across (i, j) is a logistic cost function, which is to be minimized with respect to the f^{ij} for a given set of headways (the operational problem). The f^{ij} 's must be in the unit interval; they are not restricted by other constraints. Therefore, since the objective function is separable in the f^{ij}, each f^{ij} can be chosen independently of the others, by minimizing its contribution to the objective function. In physical terms this means that for a given set of headways, we will ship without transshipments if the direct-cost per item, z^{ij}, is smaller than the marginal cost of sending an item through the terminal:

$$\frac{2c_s + \left[r^i + r^j \right] c_d}{v_{max}} + c_h \max\left\{ H^i, H^j \right\}.$$

Otherwise, we should ship through the terminal. Because f^{ij} should be either 0 or 1, all the sources in an origin subregion will ship in the same manner.

If the system is operated on the clock with a unique headway H, and the conditions are fairly homogeneous – so that z^{ij} is the sum of a constant and $r^{ij} c_d / v_{max}$, where r^{ij} is the distance from i to j – then the decision for f^{ij} is only based on the circuity of the terminal route. We ship direct whenever $r^i + r^j - r^{ij}$ is greater than a fixed quantity.

If the vehicles cannot be operated full, then the operational problem is more complicated because the f^{ij} 's don't define a separable linear program anymore, but it should be possible to solve it approximately with relaxation schemes. Note, however, that with the high flows likely to arise in this type of problem (otherwise we would not be considering the direct shipping option to begin with) it is quite unlikely that route length constraints or pipeline inventory cost considerations would prevent filling the vehicles.

The tactical problem is easy to solve if the operational problem can be solved. If the system is operated on the clock (with a unique H), it is a simple matter to choose the best H by testing various values; the one with the least cost should be chosen. If a unique headway is not desirable the solution is more difficult. We have seen, however, that conditions have to be drastically different for some i's and j's for that to be the case. It should then be fairly obvious which (i,j)'s should be operated on lower or higher frequencies than the norm; and to identify a few reasonable headway sets for testing. These techniques can be applied to Problem 6.4.

More detailed discrimination: Decomposition approaches also work if other details of the origins and destinations, besides location, are also used to discriminate across O-D pairs. This is desirable if the additional discriminating characteristics, such as production rates and item values, change drastically across origins and destinations. In this case, instead of partitioning the region into subregions, one must deal individually with the specific origins and destinations; but this detailed approach is not difficult to apply if vehicles are not allowed to either peddle or collect.

For this scenario, Hall (1987) bases routing decisions on both location and the specific demand and production rates. He assumes that the total logistic cost of the links inbound to the terminal is linear in flow. (We have seen already that this will happen if these links are operated with full vehicles; thus the assumption implicitly assumes that the flows on these links are fairly substantial). With this assumption, the problem decomposes by destination: the best way of receiving items at each j from all the i's can be determined independently for all j's. Each of these destination subproblems can be further decomposed if one holds constant the headway (or the shipment size as Hall suggests) from the terminal to the destination. The method can be applied with and without coordination at the terminal; and can probably be extended to allow for multi-stop non-terminal routes.

Blumenfeld et. al (1985a) have addressed the same problem without the linearity condition, but their method only works for few (3 or 4) origins and any number of destinations. The problem is also decomposed by destination; but to achieve independence one must condition jointly on the inbound headways (or shipment sizes) from every origin to the terminal.

It is impossible to explain here all the possible situations and how they could be addressed (see Hall, 1993b, for more examples). Suffice it to say that if some sort of asymmetric service is suspected to be beneficial it might be possible to use a decomposition method if a proper set of conditioning (tactical) variables can be found.

6.4 Multi-Terminal Systems: One Transshipment

The remainder of this chapter discusses systems with more than one terminal, and for the most part it will address strategic problems. We seek the location and number of terminals that should be operated, as well as the schedules and routes to be used. This section considers the case where each item is transshipped at most once, and section 6.5 multiple transshipments. Symmetric strategies will be examined in some detail, with discriminating customer treatments discussed only briefly. We begin with an extended discussion of the operational problem – that of determining the vehicle and item routes for given terminal locations and dis-patching frequency – as it is of central importance with multiple terminals.

6.4.1 The Operational Problem

A building block toward tactical and strategic analyses, the solution to the operational problem is also of intrinsic interest to public carriers. Because public carriers do not haul their own freight, they cannot determine precisely the value of the items moving through their system and the ensuing inventory costs. Thus, for these carriers the tactical problem is somewhat academic. In practice the service level (e.g.daily deliveries) is chosen based on marketing considerations, and is widely advertised. The market then determines which types of commodities move through the system.

The routing schemes about to be introduced extend those in Hall (1984), Hall and Daganzo (1984), Daganzo (1987c), and Campbell (1990b). For clarity, they are described for a one-dimensional region first, with 2-dimensional generalizations introduced later. For the one-dimensional case we describe non-hierarchical solutions – where the same flow is routed through all the terminals – first, and more efficient hierarchical methods second.

Non-hierarchical routing on the line: Figure 6.6a displays a region **R** and $N_T = 7$ evenly spaced terminals. We assume that there are many origins and destinations in the region (N_o, $N_d \gg N_T$.)

The non-redundancy principle introduced in Chapter 5 for one-to-many networks also applies here; with only one transshipment allowed, the flow between each O-D pair should move through only one terminal. As a result each terminal has a separate set of origin-destination pairs to serve. Given this set, each terminal should be operated as studied in the previous section. We will assume (reasonably so – see problem 6.1) that all the origins and destinations are served with the same headway H. As a result the number of stops made by vehicles on their peddling and collecting routes must be adjusted by location in response to the spatially changing demand and supply rates.

The terminal could then be operated on a clock, with all the vehicles arriving and leaving the terminal at once, for minimal delays to the items. We will also assume that H is the same for all terminals. This is reasonable because a unique H simplifies the operating plan and the job of advertising the service schedules.

The best operating plan will minimize the total vehicle-miles and the number of vehicle stops. We assume that pipeline inventory can be neglected, and that vehicles leave and arrive at the terminals full.

Assuming that each origin generates less than a truckload of goods per headway, the goods it ships through a terminal can be collected with a single stop by a single collecting vehicle. Hence, the number of collection stops made during H at one origin, m^o , equals the number of terminals to which that origin is shipping. Similarly, the number of delivery stops per destination, m^d , is the number of terminals from which deliveries are received. Thus, the number of stops made in H is a direct function of the allocation of O-D pairs to terminals. In a subregion (interval) of unit size (length) the number of stops is: $\delta^o m^o + \delta^d m^d$.

As in Sec. 4.2, we define collection (distribution) *line-haul* distance of a terminal as the average distance to (from) the terminal from (to) every origin (destination) using it, multiplied by the number of collection (distribution) tours started at the terminal. *In other words, the total line-haul distance in R equals the number of item-miles traveled, divided by v_{max}. Therefore, it is uniquely defined by the allocation of O-D pairs to terminals.*

Note that if each vehicle were to make only one collection stop, then the line-haul distance would equal the total distance traveled. Because vehicles make multiple stops, the total distance traveled is greater than the line-haul distance. In agreement with Sec. 4.2, we call the distance added by the stops "local distance". We now show that the local collection distance traveled per headway in a given region is proportional to the number of stops made in the region, except for a constant that can be ignored. First note that the local distance for a tour with n_s stops is: $(n_s-1)/(2\delta^o)$. This is true because for every *two* stops added to a tour, its length only increases by *one* interstop distance, $(\delta^o)^{-1}$. (You must imagine that one stop is tacked on to the end of the tour and the other one to the beginning, so as to keep the tour's center of gravity fixed; then only the stop at the far end lengthens the tour.) Clearly, according to the formula, each collection stop made in a region contributes $(2\delta^o)^{-1}$ distance units to the total local distance, and each vehicle tour subtracts the same amount from this total. Because the total number of collection tours is fixed (remember that vehicles travel full) the total distance deducted in R is a constant, which we ignore here. The same occurs for distribution, where each stop adds $(2\delta^d)^{-1}$ distance units. This establishes that the local travel costs, for both collection and distribution, only depend on the number of stops made in the region; the less the better.

Since the number of stops is a direct function of the O-D allocation to terminals, an allocation uniquely defines the local travel costs; as well as all the stopping and line-haul travel costs.

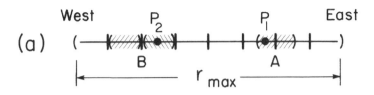

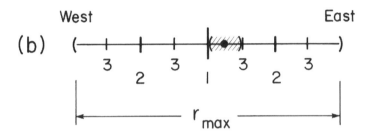

Fig. 6.6   Standard vs. hierarchical routing and terminal systems

In the following we examine various allocation strategies and their effect on stops, local distance and line-haul distance. Since local distance is proportional to the number of stops, to assess the efficacy of an operating strategy, it suffices to keep track of the line-haul miles and the number of vehicle stops.

A possible strategy is depicted by the arrangement in Fig. 6.6a. The shaded area around terminal "A" represents its collection influence area. We assume that all the items from the shaded area are shipped through "A", regardless of destination, and that origins outside the shaded area ship through other terminals. Consequently, from that terminal items are delivered to all destinations. If all the terminals in Fig. 6.6 (or in the two-dimensional case depicted in Fig. 6.7) were to operate in this manner the influence areas would partition the service region and, we would have: $m^o = 1$ and $m^d = N_T$; thus, the number of stops per headway, per unit length would be: $\delta^o + \delta^d N_T$. If there are more destinations than origins then it would be better to define *distribution* influence areas and the number of stops would be smaller: $\delta^o N_T + \delta^d$; we will assume without loss of generality that this is not the case.

The strategy we have just described is termed 1-terminal routing because each (small) area either ships or receives from only one terminal. A drawback of the strategy is that items sometimes travel more line-haul miles than the minimum possible, as happens for an item traveling from P_1 to P_2 in Fig. 6.6a.

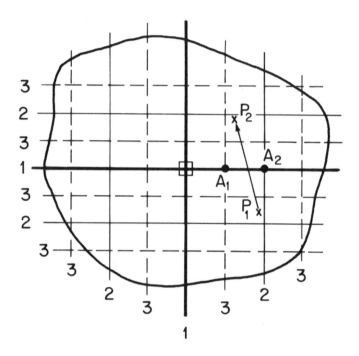

Fig. 6.7   Hierarchical routing in two dimensions

An alternative routing scheme that eliminates this back-tracking is illustrated with terminal "B". This terminal has two influence areas, displayed by the cross-hatched segments to its right and left, but it only draws part of the supply from these areas. The influence area located to the left of B ships through B only items destined for points east of B (as well as for 1/2 of the points in the influence area that are closer to B than to any other terminal.) The influence area located to the right, similarly, sends items to all points west, and to the points within itself that are closest to B.

This 2-terminal routing scheme eliminates back-tracking for most origin-destination pairs, except for O-D pairs lying entirely within two neighboring terminals. The ensuing savings in line-haul distance are achieved at the expense of one extra stop per origin. Since $m^o = 2$, the number of stops per headway and per unit length is now: $2\delta^o + \delta^d\, N_T$.

In going from 1- to 2-terminal routing we save approximately $r_{max}/4N_T$ line-haul vehicle-miles per inbound vehicle tour (since in the first case about 1/2 of the inbound miles are backtracking miles and in the second case nearly none), but we add δ^o stops per unit length per headway. If the level of demand is such that we require N_v vehicle tours to collect all the items in **R** during a headway ($N_v = \lambda \, r_{max}^2 \, H/v_{max}$) then the saved line-haul vehicle-miles per headway in **R** are: $r_{max}N_v/4N_T$. Usually, $N_v >> N_T$, and the total line-haul distance saved should be several times larger than r_{max} . The extra local collection distance, on the other hand, is negligible by comparison since it equals: $(\delta^o)r_{max} \, (2\delta^o)^{-1} = r_{max}/2$. Thus, only if the intrinsic cost of a stop, c_s, is large enough to nullify the line-haul savings, would the 1-terminal strategy be preferable. This is a moot issue, however, because the number of stops can be reduced below the 1-terminal levels, *without additional backtracking*, by hierarchical schemes.

Hierarchical routing on the line: So far, as in Hall (1984) and Hall and Daganzo (1984), terminals have not been differentiated in any manner; if the origin and destination flows don't change much with location, the flow passing through each terminal is nearly the same. These strategies, however, result in many more delivery than collection stops, or the opposite. We illustrate now how one can greatly decrease the number of delivery stops with a small increase in the number of collection stops. We will do this first for a one-dimensional region (as in Fig. 6.6) and later for a two-dimensional region (as in Fig. 6.7).

Figure 6.6b shows the same region and terminals of Fig. 6.6a, but now the terminals have been labeled by numbers. The terminal near the center is labelled "1"; it partitions **R** into two equal halves. The terminals located near the middle of each half are labelled "2", and the ones located near the middle of each fourth are labelled "3". These labels represent levels within a hierarchy of terminals, with "1" being the highest level. A system with a full set of terminals and $\ell = 1, ... , L$ levels will have $2^L - 1$ terminals and $2^{(\ell-1)}$ terminals at each level. (In earlier chapters L denoted the number of time periods, but that variable is not used in this chapter.)

A routing strategy that avoids backtracking could be defined as follows: Serve each O-D pair by the highest level terminal between the origin and the destination; if there is no terminal in between, use the neighboring one which can be reached from both customers with the least combined distance.

The definition uniquely identifies a terminal for each origin-destination pair; *because there can never be a tie for the highest level between terminals.* This is true because, with our labelling strategy, two terminals of the same level are always separated by one or more higher level terminals.

Notice that each origin sends items exactly through L terminals and receives items through L terminals. For example, origins in the cross-hatched section of Fig. 6.6b would ship through the level-1 terminal for all points west of the section, through the level-3 terminal at the right end of the section for points in the neighboring section to the right, and through the level-2 terminal on the right half of the region for the remaining points farther east. The destinations in the shaded region would also receive items from the whole region through the same three terminals. (It is recommended at this point to identify mentally the 3 terminals that would be used for each of the 8 segments in the figure). Not given in detail here, a formal proof of our statement for arbitrary L can be constructed along the following lines. For any origin segment, one would start by identifying the set of destinations served through the level-1 terminal. Recognizing that lower-level terminals within this set are not used, one would then show that only one of the level-2 terminals is used to serve the remaining points. The argument would then be repeated for lower levels. Thus, without increasing line-haul (backtracking) miles, the number of stops can be reduced to $L(\delta^o + \delta^d)$ from $2\delta^o + \delta^d N_T$. If $\delta^o \approx \delta^d$, the reduction in the number of stops can be quite substantial: from $m^o + m^d \approx N_T$ to $m^o + m^d \approx 2L = 2\log_2(N_T+1)$. Thus, we see that with a hierarchical routing strategy, the number of stops only increases logarithmically with the number of terminals.

Note that the flows through the various terminals are radically different even if the origins and destinations are uniformly distributed. Not counting O-D pairs within a segment, the level-1 terminal handles 1/2 of all the origin destination pairs. The level-2 terminals handle 1/2 of the rest; i.e. 1/4 of the total. Since there are two level-2 terminals, each handles 1/8 of the total. Assuming that there are more than 3 levels, the level-3 terminal would handle 1/2 of the rest: $(1/2)^\ell$ (for $\ell=3$), and since there are $2^{(\ell-1)}$ terminals of this type, each would handle $2/(4^\ell)$ of the traffic, etc...

Hierarchical terminal systems are used by many common carriers. Federal Express, an overnight package delivery carrier (H=1 day), started their operation with one hub in Memphis (Tennessee) and later opened another hub in Oakland(California). The Oakland hub is a secondary hub that only serves O-D pairs in the Western United States, and is in our terminology a level-2 terminal. Federal Express operates nowadays with L=2. Similar hierarchies can be found upon inspection of airline networks, although in that case the highest level terminals cannot carry as much traffic as it would be ideal because of airport capacity limitations.

Two-dimensional extension: We are now ready to see how the hierarchical strategy can be extended to two dimensions. In this case it helps to think in terms of two sets of parallel lines in two perpendicular directions, defining a square grid as shown in Figure 6.7. Each set of lines is numbered with the bisecting strategy used in Fig. 6.6b, with the result shown in Fig. 6.7. The dark (level-1) lines should cross near the center of the region, **R**, and terminals are assumed to be located at or near the intersection of any two lines (level-3 lines are represented by dashed lines in our figure). Thus, with L=3, there should be a maximum of $(2^\ell-1)^2$ terminals, since there are $(2^\ell-1)$ lines in each family. The actual number of terminals may be smaller if some of the lines intersect outside **R** .

The terminal selection process for a given O-D pair is simple. Choose the highest numbered line from each set that is crossed by the segment joining the origin and the destination, and use the terminal located at the point of intersection.

As in one dimension, this defines unambiguously the terminal to be used, unless the trip does not cross a line in one of the directions. If this happens, one is assumed to choose the least circuitous terminal on the highest level line crossed in the other direction; thus, in traveling from P_1 to P_2 an item would be shipped either through A_1 or A_2; see Figure 6.7. If the path crosses no lines, then the origin-destination pair lies entirely within a cell of the grid and one would choose among the four terminals on the corners.

If travel were only possible in the directions of the grid (distances follow an L_1 metric) then only the trips in which one (or both) of the families are not crossed would entail some back-tracking. If the origins and destinations are independently distributed of each other, and $N_T \gg 1$, then the probability that a trip requires some backtracking in one direction will be on the order of $(1/N_T)^{1/2}$, the reciprocal of the number of lines in one direction; and the average distance added to the trip will be about 1/3 of the separation between terminals, $(|\mathbf{R}|/N_T)^{1/2}$. This extra distance result holds because: (i) the sum of the distances to the best line (of the two possible) is 2/3 of the distance separation between lines (see problem 6.5), and (ii) because as is well known (e.g., Larson and Odoni, 1981), the average separation between points is 1/3 of the lattice spacing. Thus, the expected added distance is one third of the lattice spacing, as claimed; and if one considers both directions the incremental distance should be twice as large. Since the probability of backtracking in either direction is $1/N_T$, the expected added distance across all O-D pairs should then be:

$$2|\mathbf{R}|^{1/2}/(3N_T) . \tag{6.4}$$

This expression assumes that travel takes place along a grid. If this is not the case the distance added by the terminal stopover will be a different expression, but should behave qualitatively similarly.[5]

If **R** is not close to a square, or N_T is not close to $(2^L-1)^2$ for some integer L, then some levels may have less than a full complement of terminals. Figure 6.8 illustrates this; with L=3, it has N_T= 18 terminals, when N_T should have been 49. It should be clear from the derivation, however, that Eq.(6.4) should be fairly accurate, even without a full complement of terminals.

Let us now turn our attention to the number of stops per origin and destination, m^o and m^d. With a full complement of terminals, each origin would ship through L^2 terminals, and nearby destinations would receive through the same terminals; thus, $m^o \approx m^d \approx L^2$. With less than a full complement of terminals, the number of stops would be smaller. Figure 6.8 depicts the number of stops for collection (or delivery) that are made per origin (or destination) in each cell. Note that, even though L=3, only 4 cells require 9 stops. The average across cells is significantly smaller, approximately 6.7 stops.

A reasonable approximation for the number of stops is given by L^2, using for L the real solution of $N_T = (2^L-1)^2$: $L = \log_2(1 + N_T^{1/2})$. That is:

$$m^o = m^d \approx \left[\log_2\!\left(1 + N_T^{1/2}\right) \right]^2 . \tag{6.5a}$$

For $N_T = 18$, as in the figure, this yields a better approximation than using L=3; i.e., 5.7 , instead of 9 stops. This expression is exact if the amount in brackets is an integer, and other examples (e.g. with N=4 and N=12) reveal that it tends to under-predict the actual number of stops by about 10%. A simpler expression which is very accurate for $N_T<10^2$ is:[6]

$$m^o + m^d \approx 2.6 N_T^{1/2} \tag{6.5b}$$

and since Eqs. (6.5) tend to under predict the actual average by about 10% we will use instead:

$$m^o + m^d = 3 N_T^{1/2} . \tag{6.6}$$

---

[5] The expression used in Daganzo (1987c), $[2|R|^{1/2}/(3N_T)]\{1-(4N_T)^{-1/2}\}$, is qualitatively similar to Eq. (6.4). Developed by Hall and Daganzo (1984) for a 4-terminal routing strategy (which is inefficient in terms of number of stops but yields the same backtracking distance), this expression is exact when **R** is a square; it accounts for the peculiar edge and corner zones, which is only important if N_T is small.

[6] The difference between (6.5b) and twice (6.5a) is less than 1 stop, and for $N_T>3$, less than 3.6%.

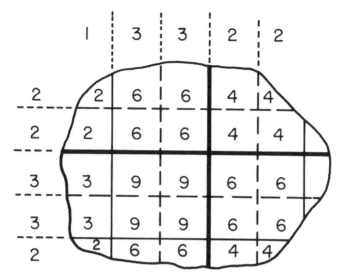

Fig. 6.8   Hierarchical routing without a full complement of terminals

We notice that the circuity distance (6.4) decreases with N_T , but the cost caused by local stops (6.6) increases with N_T . Thus, we shall look for the number of terminals that minimizes cost. Before we address this strategic problem, one last point needs to be discussed.

Detailed solution: It has been assumed so far that terminals were more or less located on a square lattice within the service region. We then showed how it was possible to develop a labeling system that minimized the number of terminals serving each point in **R** while keeping backtracking at a minimum.

If the terminal locations are given and they do not remotely resemble a lattice, one can achieve the same goal with a detailed trip assignment scheme. Essentially, each O-D pair (i , j) must be assigned to one terminal, k=1 ,..., N_T , minimizing the line-haul distance cost and, the local motion cost--including local distance and stops.

The solution can be specified in terms of zero-one decision variables x_k^{ij}, taking the value 1 if terminal k is used between i and j , and 0 otherwise. The line-haul cost equals, as before, the item-miles traveled multiplied by c_d/v_{max} , and each local stop adds $\alpha_2 \approx (c_s + c_d k \delta^{1/2})$ to the local motion cost. (Note that the distance arguments given for one-dimensional problems were equivalent to using k=0.5 in the expression for α_2). Letting r_k^{ij} denote the known distance of a trip from i to j , passing through terminal k, and D^{ij} the number of items that must be carried from i to j during one headway, we can write:

$$Total \# stops = \sum_{ik} min\left\{\sum_j x_k^{ij};1\right\} + \sum_{jk} min\left\{\sum_i x_k^{ij};1\right\}$$

and

$$Total \# Item\text{-}Miles = \sum_{ijk} D^{ij}\, r_k^{ij}\, x_k^{ij}.$$

Assuming for simplicity that α_2 is the same for pickups and deliveries ($\delta^o \approx \delta^d$), then we would like to minimize:

$$\alpha_2\left[\sum_{ik} min\left\{\sum_j x_k^{ij};1\right\} + \sum_{jk} min\left\{\sum_i x_k^{ij};1\right\}\right] + \left(\frac{c_d}{v_{max}}\right)\sum_{ijk} D^{ij} r_k^{ij} x_k^{ij} \quad (6.7)$$

where the x_k^{ij} are zero-one variables with:[7]

$$\sum_k x_k^{ij} = 1.$$

Detailed data are needed to solve this non-convex minimization problem; the procedure may be unwieldy with many origins and destinations. The two-step process espoused in this monograph can then be useful. First, pretend that the region **R** has been squeezed and stretched, as if it was made of a pliable material, until the terminals fall on a grid as in Fig. 6.8; trips can then be assigned to terminals as previously described. The resulting assignment is likely to be quite good on its own merits; and it can also be used as an initial solution in the minimization of (6.7), perhaps with simulated annealing, if the required detailed data are available.

6.4.2 Strategic and Tactical Problems

To solve the strategic and tactical problems we use the optimal solution of the idealized operational problem. This is reasonable, since at this level one should not plan to use a poor set of terminal locations. The collection

---

[7] It is not difficult to include terminal flow restrictions (e.g. requiring that $\sum_{ij} D^{ij} x_k^{ij}$ remains below some limit for terminal k) in the formulation, but this may complicate the solution procedure.

cost per item transported due to the number of stops is based on Eq. (6.6). Recognizing that the number of stops at each origin is $m^o = 1.5\,N_T^{1/2}$, we can write:

$$\begin{pmatrix}\text{inbound stop}\\ \text{cost per item}\end{pmatrix} = \alpha_2 \left[\frac{\delta^0}{\lambda H |R|}\right]\left(1.5 N_T^{1/2}\right)$$

where the quantity in brackets represents the reciprocal of the number of items collected at the average origin in one headway. A similar expression holds for the outbound stop cost.

If we assume that $\delta^o = \delta^d = \delta$ (the reader can generalize this assumption easily) we obtain:

$$stop\ cost\ per\ item = \frac{3\alpha_2\delta}{\lambda H |R|}\left(N_T^{1/2}\right) \tag{6.8a}$$

We will also assume in our exposition that δ and λ do not change over the region. If they do it is better to work with a (less intuitive) total regional cost per day; the resulting expressions, presented in Sec. 6.5.3., are close to the ones with constant conditions when averaged across items. Qualitatively similar conclusions are reached.

A line-haul circuity cost per item can be obtained from the extra distance traveled by each item (6.4). Since items travel in full vehicles, the prorated circuity cost per item is:

$$circuity\ cost\ per\ item = \left(\frac{c_d}{v_{max}}\right)\left(\frac{2|\mathbf{R}|^{1/2}}{3 N_T}\right). \tag{6.8b}$$

This cost is paid in addition to the basic line-haul cost, which is proportional to the average distance between origins and destinations; this basic cost is of order $[c_d/v_{max}]|\mathbf{R}|^{1/2}$.

We should also include a fixed cost of operating a terminal; i.e., an expression like Eq. (5.3) of Chapter 5, with $|\mathbf{R}|/N_T$ instead of I, which should increase linearly with N_T:

$$terminal\ cost\ per\ item = \alpha_5 + \alpha_6 \frac{N_T}{|R|}. \tag{6.8c}$$

Finally, we must also include the stationary holding cost at the origins and destinations:

$$holding\ cost\ per\ item = c_h H .\qquad(6.8d)$$

The sum of Eqs. (6.8) is our logistic cost function. With it we can answer a variety of questions. A strategic level question could be: how many terminals should be operated, given H ? This might be appropriate for a carrier that is planning entry in a market niche with a well defined H .

Alternatively we may be interested in determining the best H for a given N_T , or in selecting both together. Everything is possible, and easy to do, since the objective function is defined in terms of only one or two decision variables, and is unimodal (it is a "positive" polynomial of the form used in geometric programming).

For a given N_T , the best H balances local stop costs and holding cost; circuity and terminal costs are fixed. We find:

$$H^* \approx \left[\frac{3\alpha_2 \delta}{\lambda c_h |\boldsymbol{R}|} N_T^{\frac{1}{2}} \right]^{\frac{1}{2}}\qquad(6.9a)$$

and the total cost per item, not including the fixed basic line-haul cost, is:

$$\frac{cost}{item} = \alpha_5 + \alpha_6 \frac{N_T}{|\boldsymbol{R}|} + \left(\frac{c_d}{v_{max}} \right) \left(\frac{2|\boldsymbol{R}|^{\frac{1}{2}}}{3N_T} \right) + 2 \left(\frac{3\alpha_2 \delta c_h}{\lambda |\boldsymbol{R}|} N_T^{\frac{1}{2}} \right)^{\frac{1}{2}}$$

$$= \alpha_5 + \alpha_6 \frac{N_T}{|\boldsymbol{R}|} + \left[\frac{c_d |\boldsymbol{R}|^{\frac{1}{2}}}{v_{max}} \right] \left\{ \frac{2}{3N_T} + 2 \left(\frac{3\alpha_2 \delta c_h v_{max}^2}{\lambda |\boldsymbol{R}| c_d^2} \right)^{\frac{1}{2}} N_T^{\frac{1}{4}} \right\}.$$

As an example, we find the optimal number of terminals for a case where fixed terminal costs can be neglected and where the cost of a stop c_s is small compared to the distance component $k c_d \delta^{-1/2}$; thus $\alpha_6 \approx 0$ and $\alpha_2 \approx c_d k \delta^{-1/2}$. The cost per item (using $(3k^2)^{1/2} \approx 1$ and disregarding the constant α_5) is:

$$\frac{cost}{item} \approx 2 \left[\frac{c_d |\boldsymbol{R}|^{\frac{1}{2}}}{v_{max}} \right] \left\{ \frac{1}{3N_T} + \frac{K}{N_0} N_T^{\frac{1}{4}} \right\} \tag{6.9b}$$

where N_0 is the number of origins (and destinations), and K is the dimensionless constant introduced at the outset of this Chapter in connection with Eq. (6.2). The factor in brackets, comparable with the basic line-haul cost, represents the cost of crossing the service region (if it was "round" in shape) prorated to the items in a full vehicle.

The minimum of (6.9b) is obtained for

$$N_T^* \approx \left(\frac{4N_0}{3K} \right)^{\frac{4}{5}}, \tag{6.10a}$$

and the result is

$$\frac{cost}{item} = \left[\frac{c_d |\boldsymbol{R}|^{\frac{1}{2}}}{v_{max}} \right] \left\{ 2.6 \left(\frac{K}{N_0} \right)^{\frac{4}{5}} \right\}. \tag{6.10b}$$

The cost without transshipments (6.3b), when expressed as a function of the same variables, adopts the same form but the term in braces is of order $K/(N_0^{1/2})$. Clearly, if N_0 is large, terminals reduce cost dramatically.[8]

Interpretation: A comparison of the number of stops of each vehicle route is interesting. The numbers of local stops without transshipments is $n_s \approx K$. With transshipments, the number of stops is larger for the vehicles serving the lowest level terminals, located at the intersection of dashed lines in Figs. 6.7 and 6.8. Thus, we focus on these. Vehicles based at one such terminal serve all the destinations in the 4 cells next to it and no destinations beyond; a total of $4\delta^d |\boldsymbol{R}|/N_T$ customers. A similar expression holds for origins. The item flow passing through the terminal is the average flow for one O-D pair, $\lambda/(\delta^o\delta^d)$, multiplied by the number of pairs served through it. The terminal can only serve O-D pairs entirely within a square 4-cell sub-region centered at the terminal. There are $16\delta^o\delta^d [|\boldsymbol{R}|/N_T]^2$ such O-D pairs. Some of these, however, are better served by

---

[8] Equation (6.10b) also yields a smaller cost than the non-hierarchical routing method in Daganzo (1987a) when N_0 is large. The expression there is analogous to (6.10b) but the value in braces is: $1.7 (K/N_0)^{2/3}$. The number of terminals used without a hierarchy is also considerably smaller; $N*_T \approx (1.3N_0/K)^{2/3}$, for $(N_0/K) > 10^2$.

terminals on the edge of the square subregion. Thus, the actual number served through the terminal should be somewhat smaller. The reader is encouraged to prove that only 9/16 of the O-D pairs in the sub-region are served through the terminal if origins and destinations are uniformly distributed. Therefore, we can write:

$$\left(\begin{matrix} number\ of\ O\text{-}D \\ pairs\ served \end{matrix} \right) = 9\delta^o \delta^d \left[\frac{|\mathbf{R}|}{N_T} \right]^2 ,$$

so that the flow through a lowest level terminal is about: $9\lambda[|\mathbf{R}|/N_T]^2$ items per unit time, or $(9\lambda/v_{max})$ H $(|\mathbf{R}|/N_T)^2$ delivery vehicle loads (trips) per dispatch. Since these trips must collectively stop at $4\delta^d|\mathbf{R}|/N_T$ customers, the average number of delivery stops per trip is:

$$n_s^d \approx \frac{4\delta^d\, v_{max}\, N_T}{9\lambda\,|\mathbf{R}|\,H}. \tag{6.11}$$

The collection stops are given by a similar formula.

Equations (6.11), (6.9a) and (6.10a) yield the following expression for the maximum number of delivery (or collection) stops with an optimum number of terminals:

$$\frac{n_s^*}{\sqrt{N_0}} \approx 0.3 \left(\frac{K}{N_0} \right)^{2/5}.$$

With many customers ($N_o > 100K$), this value is smaller than the average number of delivery stops with the non-hierarchical strategy in Daganzo (1987c). (The result there was $n_s^*/N_o^{1/2} = (K/N_o)^{2/3}$) . For example, if $N_o = 10^4$ and $K = 10^2$, then one should operate about 50 terminals, using something like the 3-level pattern depicted in Fig.6.7, and vehicles would make a maximum of 5 stops. The number of stops without terminals would have been much greater, $n_s = 10^2$, according to (6.3a). With a non-hierarchical strategy the *average* number of delivery stops is also close to 5 but we can only use 25 terminals. Because of the increased circuity, the cost is about 20% higher.

Qualitatively, though, the results in that reference and the improved ones presented here tell the same story. As the items become more valuable, and the origins more diffuse and small the number of terminals should be reduced, as illustrated in Fig. 6.9. Cheap bulky items can be routed through more terminals, which is logical since the circuity costs will dominate. The only difference between the hierarchical and non-hierarchical results is that the optimal system can make use of more terminals since the number of stops does not increase as rapidly with N_T.

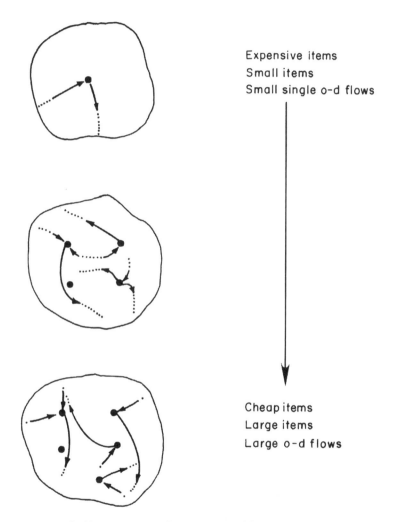

Expensive items
Small items
Small single o-d flows

Cheap items
Large items
Large o-d flows

Fig. 6.9   Desirable structures for one transshipment systems. (Source: Daganzo, 1987c)

Note also that if N_o decreases but other parameters change so as to keep K constant, e.g., the individual customers become larger, the advantage of (6.10b) over (6.3b) also decreases. If one factors in the fixed and variable terminal costs we find that the optimal N_T is smaller; not surprisingly, shipping without a transshipment eventually becomes desirable for sufficiently large customers.

6.4.3 Extensions

It was assumed until now that vehicle routes could be as long as necessary and have as many stops as needed. If this is not the case, but we are still dealing with cheap items carried in full vehicles, one can modify the optimization of the strategic and tactical problem to yield the desired result. One would still try to run the system on a clock, with a common headway, but perhaps would stop introducing new terminals as soon as the lowest level terminals resulted in routes with too many stops.

The desired system configuration would be given by the minimum of Eqs. (6.8), where N_T and H would have to satisfy n_s^d, $n_s^o \le n_{max}$, with the n_s given by Eq. (6.11). This constraint, like the inclusion of terminal costs (6.8c) in the objective function, will tend to produce a smaller N_T than suggested by Eq. (6.10a).

Equations (6.8) were developed for cheap items, and identical vehicles, but similar expressions can be developed in other cases, including situations where H can vary across terminals of different levels. Although it is impossible to cover all aspects of the problem in this monograph, it should be clear that in many cases the steps to be followed should be quite similar. The reader is encouraged to try problem 6.6, which addresses an idealized situation peculiar to airlines (the exercise extends the work of Jeng, 1987, who studied an idealized model of a single hub airline.)

If some of the origins and destinations are much larger than others it may be worthwhile to consider discriminating strategies whereby the origin-destination pairs with the largest flow would be served non-stop and the rest though the system of terminals, as in Sec. 6.3.2.

If pairs are chosen for inclusion in either one of the categories based on the amount of flow alone, with no regard for location, then it is possible to find the best O-D pair allocation (and the resulting system design) by conditioning on the number of pairs that are handled without a transshipment. For any number, the costs on the two systems are independent of actions taken to control the other system and as a result the two can be optimized separately, as we have learned previously. A near-best allocation can be formed by repeating the process for various (carefully selected) numbers of origin-destination pairs in the non-terminal system, and comparing total costs.

As with one terminal, ideally one might want to use the geographical locations of origins and destinations and relevant flow information in deciding where to allocate an O-D pair, but the problem is more complex than with only terminal. Fortunately, with several terminals the importance of location is diminished because the maximum distance added by a terminal stop-over is smaller.

In summary, this section illustrates how the number of vehicle stops and the total logistic cost can be reduced by transshipping items once at break-bulk terminals. We have seen that a hierarchy of terminals enhances the transshipment benefits.

6.5 Multi-Terminal Systems: Multiple Transshipments

This section shows how further cost reductions can be achieved with additional transshipments. Section 6.5.1 discusses two transshipments through BBTs; as in the prior sections, it will be assumed that vehicles of maximum size, v_{max}, can reach the origins and destinations. It will also be assumed that the pipeline inventory cost can be neglected, relative to transportation costs. Systems with both BBTs and CTs are examined in Section 6.5.2.

6.5.1 Two Transshipments Through BBTs

With two transshipments, a non-hierarchical arrangement of terminals is no longer asymmetric with respect to collection and distribution; as we shall see, it requires few local stops at both ends of a trip.

Thus, a hierarchy of terminal levels is not used to select an item's route through the terminals: each O-D pair is simply assigned to the least circuitous terminal pair, considering only terminals in the immediate neighborhood of the origin or the destination. In Figure 6.7 only the terminals on the four corners of the cell containing the origin (or the destination) would be potential candidates. Of the 16 possible combinations the pair adding the least distance should be chosen. A typical item would first travel to the origin terminal on a collection vehicle; it would then be sent to the destination terminal on an inter-terminal vehicle and finally, after a second transshipment, it would be delivered to its destination.

We will assume here that: (i) all the vehicles arrive and leave the terminals full, (ii) the system is operated on a clock with a common headway H, and (iii) every terminal pair is linked by a non-stop vehicle route along the shortest path; i.e., multiple stops at the terminals are not allowed. Conditions (i) and (ii), used with one transshipment systems, should be desirable here for the same reasons. Condition (iii) ensures that the vehicle-miles of inter-terminal travel are minimum. A more general set of conditions, e.g.

with different headways for collection and distribution than for local travel, does not reduce cost (Daganzo, 1987c).

With the system operated on a clock the holding cost per item is not increased by the second transshipment; it is still given by c_hH . The motion cost and terminal cost expressions are examined next.

If we decompose collection and delivery vehicle-miles into line-haul and local components as before, and define inter-terminal travel as line-haul, then the total line-haul vehicle-miles still equal the total item-miles divided by v_{max} . Consideration shows that for our routing scheme the distance added by the two transshipments is the same as the distance added with only one, Eq. (6.4). As a result, the line-haul circuity cost per item is still given by Eq. (6.8b).

Local motion costs are still proportional to the number of stops, with the same proportionality factor. As before, the number of stops per customer equals the number of terminals serving the customer in both directions. But this number, given by Eq. (6.6) for one transhipment, is now reduced to $4 + 4 = 8$. It is even less for customers along the boundary of the region. Consequently, the stop cost per item, assuming that $\delta^o \approx \delta^d$ as in Eq. (6.8a), is instead about $(3/8)(N_T)^{1/2}$ times smaller: $8\alpha_2\delta/(\lambda H|\mathbf{R}|)$.

For a given N_T and total demand, the fixed terminal costs don't change, but the terminal costs proportional to flow should just about double; after all, items are handled twice and spend twice the amount of time moving through terminals. Thus, the terminal cost per item should become: $2\alpha_5 + \alpha_6N_T/|\mathbf{R}|$.

The solutions of the strategic and tactical problems are now analogous to the minimization of (6.8). Now, however, we must introduce a flow conservation constraint. Because the system operates on a common schedule,[9] vehicles are full, and multiple stops at terminals are not allowed, the number of collection tours arriving at a terminal (i.e. vehicle loads) must equal N_T . To see this note that the number of vehicle loads collected at a terminal for other terminals must equal the number departing for other terminals, and this number is N_T-1 . Because $1/N_T$ of the freight collected is local, the number of vehicle loads collected must be N_T . The same occurs for distribution, but with $\delta^o = \delta^d$ and spatially homogeneous demand the condition is redundant. Thus:

$$\frac{\lambda |\mathbf{R}|^2 H}{N_T v_{max}} = N_T .$$

---

[9] If this constraint is relaxed, e.g. by allowing different headways, one can prove that cost can be reduced by changing one of the headways.

If terminal costs are neglected, and we use $k = 0.5$ in the expression for α_2, the optimal number of terminals and number of stops can still be expressed as a function of N_o and K alone. For $N_o > 10^2 K$ the following closed form expressions are obtained:

$$N_T^* \approx (8N_0)^{1/4} \left(\frac{N_0}{K} \right)^{1/2} \tag{6.12a}$$

$$\frac{n_s^*}{\sqrt{N_o}} \approx \frac{2^{1/2} K}{N_o} . \tag{6.12b}$$

For smaller N_o, the results are given in Table II of Daganzo (1987c). The results in this reference use a slightly lower estimate for circuity cost than (6.8b) but this has no noticeable numerical impact on the final result.

The total cost per item also assumes a form similar to Eq. (6.10b). But now (for all N_o) the factor in braces is even smaller; it is $(2.8K/N_o)$. The difference between the two factors reaches a maximum, about 0.15, for N_o/K on the order of 10^1. But the actual difference is smaller because in deriving the one transshipment results we used two conservative simplifications, Eqs. (6.4) and (6.6), which lead to slightly higher cost estimates for low N_o/K. For N_o/K comparable with 10^1, the cost overestimation is on the order of 0.08. Thus, the maximum difference between 1-transhipment and 2-transshipment costs should be on the order of 7% (and not 15%) of the cost of driving an item across the service region in a full vehicle.

Let the cost of transhipping a vehicle load including fixed delays and handling cost, $\alpha_5 v_{max}$, be momentarily defined in terms of the cost of driving a vehicle a critical distance, $c_d r_{crit} = \alpha_5 v_{max}$. Then, adding one transshipment to every item would have the equivalent effect of adding r_{crit} miles to the distance traveled by each item in a full vehicle. Clearly, two transshipments should not be considered if $r_{crit} > 7\%$ of the diameter of the service region. For $r_{crit} \approx 10^2$ miles (a value typical of trucking operations) only service regions as large as the largest countries in the world have the potential for benefiting from two BBT transshipments.

6.5.2 Many-To-Many Systems with Consolidation Terminals

The above statements do not imply that items (e.g. a letter) should not be handled more than twice between an origin and a destination; we are only stating that there is no need to have them pass through more than 1 or 2 *break-bulk* terminals – terminals serving multiple origins *and* multiple destinations where vehicles of similar characteristics swap their loads.

Consolidation terminals (connected with either a single origin or a single destination) should also be used to achieve the two main functions described in Chapter 5: reducing the length of delivery and collection routes, and allowing small vehicles to reach the origins and destinations.

A rationally designed many-to-many system might be organized as shown in Fig. 6.10, using both consolidation (CT) and break-bulk (BBT) terminals. Each consolidation terminal would collect (and distribute) items from origins (and destinations) in an influence area around it; influence areas would form a partition of the service region to ensure that service is provided everywhere. Conceivably one could have smaller CT's within each of these influence areas, but this is unlikely. The upper level CT's, shown by dots in the figure, would then become the entry points in the many-to-many network of BBT's, shown by squares in the figure. The figure denotes by arrows the paths that items either originating or ending in influence area 1 would take on the network; a single BBT transshipment is assumed. Conceivably, the BBT's themselves could also be gates to the system, acting like upper level CT's with their own influence areas.

A conditional decomposition approach, combining the result of Chapters 5 and 6, can be used to develop desirable structures for an integrated logistic system such as the one in the figure. Conditional on the size, I_{CT}, of the influence areas of the consolidation terminals, i.e. on the number of gates (e.g. post offices) to the break-bulk network, $N_o = N_d = |\mathbf{R}|/I_{CT}$, it is possible to determine the near-minimum cost per item on both portions of the system.[10] On the consolidation portion of the system within the influence areas, one ould use the methods of Chapter 5, and on the break-bulk portion those of Secs. 6.2, 3 and 4.[11] In addition to N_o, we may want to freeze N_T and the headway for the BBT network, H. In this way we can conveniently explore the economic merits of synchronizing the operations on both networks, and can invoke the results of either Sec. 6.2, 6.3, 6.4 or 6.5, depending on whether $N_T = 0$, $N_T = 1$, or $N_T \geq 1$. The values of N_o, N_T and H that minimize the sum of both costs should then be chosen. A more detailed design can then be developed as we have already learned, perhaps using fine tuning tools with detailed data. Problem 6.7 illustrates the approach.

---

[10] The discussion assumes that only the CT's act as gates to the system.

[11] If BBT's can be operated as gates, the procedure needs to be modified slightly; one should remember to recognize the terminal cost savings resulting from combining BBT and CT functions at the BBT's.

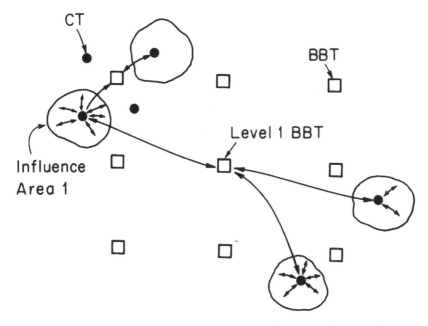

Fig. 6.10   Many-to-many system with both consolidation and break-bulk terminals

A recent application of these ideas, with some further development, is documented in Smilowitz (2001) and Smilowitz and Daganzo (2004), which describe an effort to design and evaluate large scale, integrated package delivery systems such as those of UPS and FedEx. These references examine the conditions under which it makes sense to integrate an air-express network into an existing ground network. The CA techniques proved to be practical and accurate cost predictors.

They revealed that the larger a ground network, the more efficiently it can absorb a given air network. This helps explain why UPS has chosen to run an integrated air/ground network, but not FedEx. Of course, other factors can also contribute to such decisions. Labor issues are perhaps the most obvious, since existing contracts would likely have to be renegotiated after a structural change in operations.

6.5.3 Extensions

Variable demand: We have not discussed in detail in this chapter how one should handle non-homogeneous origin and destination tables. It was assumed for the most part that origins and destinations were homogeneously distributed and that the flows between regions of comparable size was relatively independent of position. In practice, though, this is not likely to happen since population densities typically change over space.

We have seen that if some customers are much larger than others it may be better to serve them without transshipments, but we did not explore how to deal with spatial variations in demand density and customer density (we treated λ, δ^o and δ^d as constants). In prior chapters we had used the continuous approximation method to deal with such variations, and this is also possible here. While it might now appear that we would have to specify an origin-destination flow table in detail, $\lambda(x^o, x^d)$, the solution is mostly sensitive to the generation and attraction rate densities: $\lambda^o(x)$ and $\lambda^d(x)$.

With variable demand we would still try to locate the BBT terminals on a lattice, but would want to vary the number of consolidation terminals and their operation according to location. One may also wish to locate more BBT's in high density areas, but we will ignore this for the moment. The optimal solution, thus, is defined in terms of the H, N_T, $H_{CT}(x)$, and $\delta_{CT}(x)$, where $\delta_{CT}(x)$ is the spatial density of CT's in the vicinity of x, and $H_{CT}(x)$ is the headway used at those CT's. We now show that, holding H and N_T constant, the total cost decomposes locally in a manner that allows H_{CT} and δ_{CT} to be defined with the CA approach.

The motion cost during consolidation is independent of H and N_T. For a given terminal, it only depends on λ^o, λ^d, δ_{CT} and H_{CT}, and can be prorated to small sub-regions of R as a function of x alone. If no transshipments take place in the consolidation area, the average motion cost per item for collection is given by the function $z_m^o(\lambda^o, \delta o, \delta_{CT}^{-1}, H_{CT})$, defined in connection with Eq. (5.9) of Chapter 5. As stated, the arguments of z_m^o only depend on x. Similarly, the motion distribution cost is: $z_m^o(\lambda^d, \delta^d, \delta_{CT}^{-1}, H_{CT})$.

The two components of BBT motion costs are also well behaved. Given N_T and H, the circuity (line-haul) cost per unit time over R is insensitive to $\delta_{CT}(x)$ and $H_{CT}(x)$. We note that flow will pulse through BBT's differently if these variable change, but the total item-miles should remain fixed. Therefore, this cost can be ignored for the minimization of the consolidation terminal variables. Important for the minimization of N_T and H, we also note that the average circuity cost per item should be rather insensitive to $\lambda(x^o, x^d)$ if O-D trips are comparable with the diameter of R.[12] Equation (6.8b) should then be a good approximation. The BBT (local) stop cost per unit time and unit area arising from visits to the CT's can be obtained simply. Equation (6.6) gives the number of stops at each CT every headway.

---

[12] If there were predominant directions of travel along well defined corridors – stable with time – circuity might be reduced by positioning the lattice on favorable locations. However, the resulting savings (i.e. the solution's dependence on $\lambda(x^o, x^d)$) should be rather mild in most cases.

Since there are δ_{CT} CT's per unit area and each stop costs $\alpha'_2 = c_s + c_d k \delta_{CT}^{-1/2}$, the BBT system's local stop cost per unit time and unit area, z'_s is:

$$z'_s\left(\delta_{CT}, N_T, H\right) \approx 3\left(\frac{N_T^{\frac{1}{2}}}{H}\right)\left(\alpha'_2 \delta_{CT}\right).$$

We have argued in Chapter 5 that the terminals' cost per unit time should be of the form α_5(# items/unit time) + α_6(# terminals) . This total cost for the CT's can be prorated locally to small areas as a cost per unit time and unit area, z'_T:

$$z'_T\left(\lambda^o, \lambda^d, \delta_{CT}\right) = \left(\lambda^o + \lambda^d\right)\alpha_5 + \alpha_6 \delta_{CT}.$$

Consistent with Eq. (5.3), this expression yields the total cost when integrated over $\mathbf{R}$. It is again independent of the origin-destination flow details, $\lambda(\mathbf{x}^o, \mathbf{x}^d)$.

Finally, we must account for holding cost. If the schedules of the break-bulk and consolidation vehicles are not synchronized, then the waiting cost for an item traveling from $\mathbf{x}^o$ to $\mathbf{x}^d$ is $c_h(H_{CT}(\mathbf{x}^o) + H + H_{CT}(\mathbf{x}^d))$. The total holding cost for all items can be prorated to a unit area per unit time so that it only depends on the location:

$$z'_h\left(\lambda^o, \lambda^d, H, H_{CT}\right) = \left(\lambda^o + \lambda^d\right)c_h\left[H_{CT}(\mathbf{x}^o) + H/2\right].$$

If the two systems are synchronized and $H = H_{CT}$ for all locations, then the total holding cost can also be prorated locally:

$$z'_h(\lambda^o, \lambda^d, H) = (\lambda^o + \lambda^d)c_h H/2.$$

In either case, z'_h is independent of $\lambda(\mathbf{x}^o, \mathbf{x}^d)$.

Since all the cost components can be prorated to small sections of $\mathbf{R}$ as a function of $\mathbf{x}$ alone, given N_T and H , it is possible to obtain the best H_{CT} and δ_{CT} for any location, $\mathbf{x}$, by minimizing the sum of z'_h , z'_T , z'_s , the collection motion cost $\lambda^o z_m^o(\lambda^o, \delta^o, \delta_{CT}^{-1}, H_{CT})$, and the delivery motion cost $\lambda^d z_m^o(\lambda^d, \delta^d, \delta_{CT}^{-1}, H_{CT})$. The minimum of this sum is a function only of $\mathbf{x}$, N_T and H: $z'^*(\mathbf{x}, N_T, H)$. Integrated over $\mathbf{R}$, it yields the (approximate) total system cost per unit time for a given N_T and H , exclusive of the BBT circuity costs and BBT terminal costs. Reasonable values for N_T and H can now be found easily numerically by minimizing the sum of the integral, the BBT circuity costs Eq. (6.8b) and the BBT cost (6.8c). The last two costs expressed per unit time for the whole **R** are:

$$\left[\begin{array}{c} circuity \\ cost \end{array}\right] = \left(\frac{Dc_d}{v_{max}}\right)\left(\frac{2|R|^{\frac{1}{2}}}{3N_T}\right)$$

and

$$\left[\begin{array}{c} BBT \\ terminal \\ cost \end{array}\right] = 2\,D\alpha_5 + \alpha_6 N_T,$$

where D is a constant, representing the total number of items traveling per unit time:

$$D = \int_R \int_R \lambda\left(x^o, x^d\right)dx^o dx^d .$$

We have assumed that the N_T terminals would be homogeneously distributed over **R** , but in practice one would try to locate them at the intersections of major flow corridors if these have been identified in order to reduce circuity costs. The location of the BBT's, however, does not affect any of the costs used in the above calculations, except perhaps for the line-haul circuity cost. By providing more BBTs in sections of **R** with heavy demand and higher concentrations of CT's it may be possible to reduce the extra line-haul distance traveled by an average item below $2|R|^{1/2}/(3N_T)$. Any such adjustment, however, should change the distance considerably less than a small percentage increase in N_T . Thus, even if the circuity distance with the adjustment could be quantified by a more detailed expression, the resulting optimization would likely yield a similar value for N_T and H . Given what we know about the robustness of solutions (see Chapter 3), the solution we did obtain should result in costs that are not far from the ideal.

If desired, and once the locations of the CT's have been chosen, one may be able to formulate and solve approximately a detailed optimization program similar to (6.7), with the location of the N_T break-bulk terminals as decision variables in addition to the flow allocation variables, x^{ij}_k . We are optimistic about such endeavor when the desired number of BBT's is not large as would occur when the number of gates to the BBT network is itself moderate. If the number of BBT's is large, then circuity costs are small and minor reductions to it are of secondary importance.

Backhauls: It was assumed throughout this chapter that vehicles could be usefully employed at the end of their trips. This effect was captured by adjusting the motion cost coefficient α_1. Although not exclusively, this assumption is reasonable if the origin-destination flows are balanced (i.e. $\lambda(\mathbf{x}^o, \mathbf{x}^d) \approx \lambda(\mathbf{x}^d, \mathbf{x}^o)$); then when a local vehicle finishes the last delivery it is automatically well positioned to start a collection run with little deadheading. Furthermore, with a symmetric O-D flow pattern the inbound and outbound flows at every BBT are equal, which obviates the need for interterminal empty vehicle travel.

If the demand is unbalanced, a more accurate accounting of vehicle-miles is necessary since partially empty vehicles will either be arriving or departing from the terminals. The problem is likely to be more severe for BBT's than for CT's, since vehicles travel longer distances between CT's and BBT's than between CT's and individual customers.

Models and formulas have been developed to estimate CT vehicle mileage when vehicles and crews are based at an individual terminal and vehicles backhaul between the last delivery and the first pick-up (Daganzo and Hall, 1990). Such formulas would also apply to BBT routing with one transshipment.) An extensive algorithmic literature on the VRP with backhauls also exists (Casco et al., 1988). When the imbalance between inbound and outbound freight is significant, Daganzo and Hall (1990) shows that the distance traveled is just barely greater than the distance that would have to be traveled to collect (or distribute) the dominant direction of flow only, as if the other direction did not exist. We have already pointed out in Sec. 6.3.1 that this would require doubling α_1 for the dominant direction and setting it equal to zero for the secondary. With the proper distance formula, it should not be difficult to duplicate the analysis in this chapter.

If vehicles can visit a number of terminals, and some neighboring terminals have opposite imbalances, empty miles might be reduced further by backhauling from the last delivery of one terminal to a pickup *of the neighboring one* and balancing that trip by sending an empty vehicle from the second terminal to the first. The advantage of multi-terminal backhauling is particularly clear for inter-terminal vehicles in a two transshipment system. Now too, imbalances between pairs of BBT's result in some BBT's having an excess of inbound flow and others an excess of outbound flow. The solution to the Hitchcock problem of linear programming can be used to route empty BBT vehicles among terminals to minimize empty miles. The solution to the Hitchcock problem, however, may require some crews to visit several BBT's before returning to their home base; an outcome which is not desirable in practice. Other real life constraints also complicate the decision. Since carriers can greatly benefit from fewer empty miles, substantial research efforts have been made to improve backhauling decisions (see for example: Jordan, 1982, Powell et. al, 1984, and Dejax and Crainic, 1987). Most of these works, however, are algorithmic in nature, dealing with peculiarities such as real time control with imperfect information, and don't yield simple distance formulae as a function of few descriptors. This is indeed difficult for this problem. Jordan and Burns (1984) and Hall (1990) have sought estimates of empty miles for small networks, using strategies where each vehicle visits at most 2 BBT's before returning home; see also problem 6.8. This is pessimistic, however.

We would like to estimate empty vehicle miles as a simple function of N_T, which could then be used with Eqs. (6.8) to explore the various trade-offs. A somewhat optimistic estimate of this quantity, good for large N_T, is given in Daganzo and Smilowitz (2004). This reference proved with a combination of dimensional arguments and mathematical analysis that the expected distance required to reposition an empty truck in a large homogeneous system operated with the Hitchcock recipe must be insensitive to the shape of the service region, and is given by:

$$\delta_{BBT}^{-1/2}\left(a + b \log_2 N_T\right),$$

where δ_{BBT} is the spatial density of BBT's, and a and b are dimensionless constants that depend on the metric. Simulations show that a $\approx$ 1 and b $\approx$ 0.078 in the Euclidean case. In practical terms, this means that the average distance traveled by an empty truck is roughly comparable with the separation between nearby terminals; i.e., that it is about twice as long as for the Euclidean TSP, for $N_T \approx 2^5$ to 2^{10}. By multiplying the expected distance formula with the expected number of back hauls (easy to estimate if an underlying stochastic model is given), we can estimate the total expected empty vehicle miles. The result is optimistic because it is based on the Hitchcock problem, which slightly underestimates the distances of the real world.

Fortunately, formulas for empty back hauls do not have to be very accurate, because in most cases empty miles should be a small proportion of the total, and do not depend heavily on N_T. After all, if flow imbalances are serious, a carrier will normally take marketing steps to correct the imbalances since every extra item in the non-dominant direction can be carried without extra vehicle-miles. This can be done by pricing directions differently, or by other means. For example, rental car companies have drive-away programs to reposition their fleet and, because UPS's parcel flows tend to be heavier in the westward direction, that firm has considered using their trucks for carrying California produce toward the eastern U.S.

Large scale manufacturing systems: The methods and ideas we have presented in this monograph can be extended to the organization of manufacturing systems (e.g., to the planning and design of supply chains). We saw in Chapter 5 how the factory location problem was a special case of the terminal location problem in a one-to-many distribution network. This view was premised on the assumption that the inputs to the manufacturing process were ubiquitous; i.e. that changing the locations of a factory did not change the inbound logistics costs, which then could be ignored to define the system.

If some of the inputs are not ubiquitous, and must be obtained from fixed sources regardless of location, then the one-to-many model does not hold. But, we can view the *production* process as a many-to-many *logistic* process that conveys these raw materials from their sources to destination markets, in the form of a final product, passing through factories and terminals on their way.

Factories can be viewed as special kinds of terminals which somehow change the nature of the items entering and leaving. In this monograph terminals satisfied a flow conservation equation ensuring that the number of items (e.g. tons) entering a terminal were in the long run equal to the number leaving. But this weight (or volume) conservation des not apply to factories. Burns (1986) likes to distinguish between factories that transform raw materials into parts, reducing weight and volume (production plants), from those that combine parts into bulkier final products (assembly plants).

Transshipment points in a manufacturing network must be treated differently depending on whether they are bulk reducing, bulk conserving, or bulk increasing. Depending on the industry, each item may be produced and assembled at an integrated factory in a single location, or they may not. In an integrated system several factories may manufacture the items, but every item passes through only one factory. Suitably modified, the models of Secs. 6.3 and 6.4 would apply. They would have to capture the different transportation needs of the inbound and outbound items. This has been preliminarily explored for the one-factory problem when all the vehicles arrive and leave the factory fully loaded in Bhaskharan and Daganzo (1987). This report shows that most of the logistics costs are independent of location; and that only the cost of overcoming distance (with full vehicles) depends on it. As a result, the best location is the solution to a Weber problem (already discussed), *where origins and destinations have weights which reflect the ease of transport and the value of their items.* If inbound volume greatly exceeds outbound volume then there is an incentive to locate the factory close to the raw material sources; if the factory adds much value, then there is an incentive to locate the factory near the markets.

If there are clearly defined zones for raw materials and markets, and these are far apart, then there is an incentive to "dis-integrate" the system. Production plants could be located close to the raw materials and assembly plants close to the markets. In this manner transportation costs can be greatly reduced since raw materials make their way to the market in their most easily transportable form: parts, of which waste materials have been scrapped and burned away at the source but which have not yet been assembled into awkwardly shaped final products.

For a large firm, able to operate multiple factories, it may be best to operate specialized parts plants and assembly plants. Parts with similar raw material needs would be produced in the same plant, located optimally with respect to the raw materials and recognizing the different cost of production (e.g. labor productivity and wages) at different locations. Parts would then be assembled into final products at assembly plants close to the markets, allowing production to take place where it is most efficient without incurring very large transportation costs. The practice is very prevalent in the automobile manufacturing industry, where parts are often shipped half way around the world for assembly in another country.

It seems worthwhile to extend the non-detailed methods espoused in this book to aid in a more through understanding of large scale manufacturing systems in dynamic environments. The techniques seem ideally suited to that end. They have recently been used to unveil near-optimal designs and operating rules for some simple supply network scenarios, and to quantify the difference in performance between optimally-designed centralized and decentralized networks (Daganzo, 2002, 2004). This work, however, only begins to scratch the surface of possibilities.

Suggested Exercises

6.1   A single terminal serves 10 inbound routes and 2 outbound routes. The origin-destination flow density is such the each inbound route sends 1 item to each outbound route per unit time. The average waiting cost per item for any origin destination pair belonging to inbound route, i, and outbound route, j, is proportional ($c_i=1$) to the transfer time from i to j. This transfer time is a function of the headways on these two routes, H_i and H_j, and on how they are synchronized.

The inbound and outbound motion costs per item on a specific route, ℓ, decrease with the headway on the route as: k_ℓ/H_ℓ. We assume that for the inbound routes the constants k_i are (5, 9, 3, 2, 10, 7, 8, 3, 2 and 1) and the outbound constants, k_j, are (10 and 1).

(i)   Describe what kind of operation would lead to motion costs of the form: k_ℓ/H_ℓ, and a situation that might lead to the widely different k_ℓ's used in the example,

(ii) Find the optimal headway on each route if there is no synchronization,

(iii) Find the optimal headway if the system operates with a unique, H, and obtain the total system cost per unit time,

(iv) Repeat the analysis for a synchronized system where the H_i and H_j are restricted to the powers of 2 (times a factor) as described in the chapter. Compare the results. (See Daganzo, 1990, for answer.)

6.2 Prove that if the schedules are not coordinated, for a problem with one BBT , the total tactical cost per day only differs from the tactical cost with coordination by a constant independent of the terminal position. (Hint: it is not necessary to derive an expression for the tactical cost; a succinct word proof can be given).

6.3 Five hundred suppliers, uniformly and randomly scattered over a circular region of radius 500 miles centered on the origin of coordinates, send their products to ten manufacturers through a terminal located at the origin. Each manufacturing plant requires 10 truckloads worth of goods every day, from all the suppliers combined. The locations of the suppliers change from time to time (so their coordinates are not known when the tactical and strategic decisions are made) but the manufacturing plants don't move; their locations are indicated at the end of the exercise. For the following parameter values (c_d = 2 $/veh-mile , c_s = 30 $/stop , c_h = 30 $/day-truckload) , solve the tactical level problem; i.e. determine H , and the number of stops on inbound and outbound routes; sketch as well a few of these routes. (Assume that H must be an integer multiple of a day).

Solve then the strategic level problem. It is a Weber point location problem whose objective function can be calculated, as a function of position, with a spreadsheet. Outline a 1% indifference area for the terminal location.

	Manufacturing plant coor-dinates	
j	$x_j/500$	$y_j/500$
1	-0.10	-0.74
2	0.32	-0.35
3	0.87	-0.28
4	0.36	0.11
5	-0.26	-0.93
6	-0.56	0.34
7	-0.52	0.44
8	0.51	-0.01
9	0.45	0.41
10	-0.08	0.05

6.4 An airline provides service to every city pair (i,j) in its service region either linking the city pair with non-stop service, or serving it with non-stop flights to and from the hub.

Cities are classified into K categories according to a dimensionless size parameter, p_k, normalized so that

$$\sum_k p_k = 1.$$

The trip flow from all the cities in category k_1 to all the cities in category k_2 is $Dp_{k1}p_{k2}$, where D is the total passenger flow in the system (trips per unit time). We let N_k denote the number of cities in category k, assumed to be distributed homogeneously over the region.

Assuming that all the flights shuttle back and forth between two cities (one being the hub in many instances), and that the cost of each one-way flight is $[\alpha_o + c_d(\text{distance})]$, describe a decomposition method to determine: (i) the common headway, H, for flights to/from the hub, (ii) the O-D pairs that would be served through the hub, and (iii) the frequency of service for the O-D pairs served non-stop. You may assume that there is a unique airplane type that can carry v_{max} people, that pipeline inventory costs are negligible, that the terminal costs per passenger is fixed, α_5, and that the stationary inventory cost per passenger-time unit at the origin and the destination is valued at c_i monetary units.

As an extension of this exercise, the reader with access to a personal computer may want to generate with a spreadsheet N_k random city locations – defined by cartesian coordinates with origin at the hub--for each one of the K classes, and solve the problem numerically with the spreadsheet for specific values of the data. It is recommended to use widely different values for the p_k.

6.5 Consider 2 points independently located in the unit interval when all the locations are equally likely. Let D_0 denote the sum of the distances from the two points to the left side of the interval, and D_1 the sum of the distances to the right end point. Prove that $E[\min(D_0, D_1)] = 2/3$.

[**Hint:** Note that the distances from the two points to the left end of the interval, x_1 and x_2 , are independent uniform $(0, 1)$ random variables. Sketch in the sample space for x_1 and x_2 the event $\{\min(D_0, D_1) \leq d\}$, and show that its probability is d^2 .]

6.6 **Air transportation in a Linear Country.** There are 30 equally spaced airports on a line segment that is 3,000 miles long. Each one of these airports is estimated to generate 3,000 passengers per day; one-thirtieth of this flow (1/29th if one desires to be precise) is destined to each of the other airports. That is, all origin-destination pairs have the same flow.

We assume that the airline provides a competitive level of service:

(i)   Passengers do not stop between their origin and destination more than once,

(ii) Service frequency of at least one flight per day is provided on all flight segments of the network.

Part I: Network Structure

The airline currently operates airplanes that can hold 200 passengers. We seek the minimum cost to the airline (in number of airplane-miles per passenger flown, d), and the resulting level of service to the passengers (in terms of the average additional flight miles traveled because of transfers, and the service frequency on the various types of flight segments.) Find these values for the following three routing strategies

A)   Nonstop flights provided only,

B)   Hub and spoke (one transfer hub in the middle of the country),

C)   Hub and spoke (n transfer hubs, equally spaced).

For Part C you may assume that each hub has an influence area and that those influence areas do not overlap. (The influence area of a hub is defined as the region containing all the airports that consolidate their outbound flights at the hub in question: while inbound flights at a hub arrive from within its influence area, outbound flights serve all 30 airports). Find the optimal n.

Part II: The Influence of Technology

If the cost per airplane-mile, α, is proportional to:

$$\exp\left(\frac{C}{400} - 1\right),$$

where C is the seating capacity, repeat part IC while simultaneously finding the best aircraft technology for inbound and outbound links.

Part III: Refinements

Repeat Part IC if flights are synchronized with a unique headway at each hub and we use the one dimensional hierarchical routing scheme described in the text. Determine the headway used at the different level hubs.

6.7 Equation (5.6b) of Chapter 5 defines an average cost per item for items carried from (to) a terminal to destinations (from origins) in its influence area, if as explained in connection with that example vehicles make only one stop and items are carried in full vehicles from the terminal. In that expression " I " represents the influence area around the terminal, and λ is the spatial density of trip destinations (or origins) per unit time – which we have termed in this chapter λ^d(or λ^o).

Combine Eq. (5.6b) of Chapter 5 with Eqs. (6.8) of Chapter 6 to determine the optimal number of consolidation terminals for a logistic system in which only one BBT transshipment is used, and no attempt is made to coordinate inbound and outbound schedules at the CT's . Assume that the capacity of the vehicles traveling between terminals (v_{max} in (6.8b)) is greater by a factor of 4 than the capacity of the consolidation vehicles, and that the actual number of origins and destinations is very large (much larger than the number of CT's and BBT's). Assume that CT's cost the same to hold and operate than BBT's (the coefficients of (6.8c) apply to both), and that both α_6 and c_s are negligible.

6.8 "Caca's" Foods Inc. (CFI) has two factories located in a corridor (a linear segment) of length R (see diagram). Each factory produces items of different characteristics (factory 1, cases of sake; factory 2, frozen Kobe beef). Trucks carry the goods from each factory to customers that are uniformly and closely spaced over the segment.

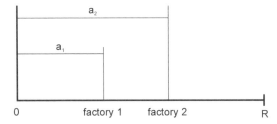

The total demand per customer from factory i (i = 1, 2) is λ_i truck loads/unit time; this is the same for all customers.

The Kuwahara Transportation Company has been hired by CFI to deliver their products in the corridor. Kuwahara's trucks can be used to transport either type of food. Devise a transportation strategy that will minimize the total vehicle-miles traveled. Neglect all inventory costs.

[**Hint:** How should you route full trucks? How about empty trucks? See Jordan and Burns, 1984, for inspiration.]

Glossary of Symbols

α_1:   Line-haul cost per vehicle trip (\$/trip),

α_2:   Motion cost per added stop/detour (\$/stop),

α'_2:   Motion cost per added stop/detour at a CT (\$/stop),

α_5:   Fixed cost per item handled at a terminal (\$/item),

α_6:   Eq.(5.3) constant, capturing terminal scale economies,

BBT: Break-bulk terminals,

c_d: Cost per vehicle-"mile",

c_h: Holding cost per item-day,

c_s: Cost per vehicle stop,

c'_s: Cost per item loaded on a vehicle,

CT: Consolidation terminal,

D: Total number of items demanded in **R** per unit time,

D':   Item flow at a point,

$\bar{D}'$:   Average item flow,

D^{ij}:   Number of items carried from i to j during one headway,

D_o: A superdestination. (Sec. 6.1 only),

δ^o(or δ^i):   Spatial density of origins (i) , (origins/area),

δ^d(or δ^j):   Spatial density of destinations (j), (destinations/area),

δ_{CT}:   Spatial density of consolidation terminals, (terminals/area),

f^{ij}:   Proportion of flow from i to j traveling directly,

H: Headway,

H*:   Optimum headway,

$\tilde{H}$:   Base headway for "power of two" strategy,

H^d(or H^j):   Destination (j) headway,

H^o(or H^i):   Origin (i) headway. (Notation is different from Ch. 5),

H_{CT}:   Consolidation terminal headway,

i: Origin label,

I: Influence area size,

I_{CT}:   Influence area of a consolidation terminal,

j: Destination label,

k: VRP local cost coefficient (dimensionless),

K: Characteristic constant of many-to-many problems,

ℓ: BBT level, for hierarchical routing,

L: Number of BBT levels for hierarchical routing,

λ (or λ^{ij}):   Origin-destination (i,j) flow density (items/time–area²). Starting in Sec. 6.2,

λ^o (or λ^i):   Origin (i) flow density (items/time-area),

λ^d (or λ^j):   Destination (j) flow density (items/time-area),

m^o: Number of terminals collecting from one origin, (number of stops per origin per headway),

m^d:	Number of terminals delivering to one destination, (number of delivery stops per destination per headway),
n_s:	Number of stops in a vehicle route,
n_s^d:	Number of destinations in a delivery vehicle route,
n_s:	Number of stops in a vehicle route,
n_s^*:	Optimal number of stops in a vehicle route,
N_o:	Number of origins,
N_d:	Number of destinations,
N_T:	Number of terminals,
N_T^*:	Optimum number of terminals,
N_v:	Number of vehicle tours,
O_o:	A superorigin (Sec. 6.1 only),
r (or r^{ij}):	Distance between origin (i) and destination (j) subregions,
r_k^{ij}:	Distance from i to j , passing through terminal k ,
r_{crit}:	Distance traveled, equivalent to one transshipment,
r_{max}:	Length of a linear service region,
r^o (or r^i):	Distance between an origin subregion (i) and a terminal,
r^d (or r^j):	Distance between destination subregion (j) and terminal,
P^i:	Area of origin subregion I,
P^j:	Area of destination subregion j,
R:	Service region,
$\|\mathbf{R}\|$:	Service region size,
t:	Time,
v:	Delivery (or pickup) lot size,
v_{max}:	Vehicle capacity (items),
x_k^{ij}:	Zero-one decision variables, indicating whether terminal k is used between i and j ,
x:	Spatial coordinates of a point,
$\mathbf{x}^o$:	Spatial coordinates of an origin,
$\mathbf{x}^d$:	Spatial coordinates of a destination,
$\mathbf{z}^i(\lambda, r, I)$:	Cost per item carried from a single origin to terminals serving influence areas of size, I ,
$z_0(\lambda, r, \delta)$:	Cost per item distributed from a single origin, without transshipments,
z^{ij}:	Average cost for items shipped direcly from i to j ,
z_m:	Motion cost per item,
z_m^o (or z_m^d):	Motion cost per item between origins (destinations) and BBT,
z_m^i:	Motion cost per item from origin i to a terminal,
z_m^j:	Motion cost per item from a terminal to destination j,
z'_s:	Local stop cost per unit time and unit area at CT's,
z'_T:	Cost of the CT's per unit time and unit area,
z'_h:	Holding cost per unit time and unit area,
$z'^*(x, N_T, H)$:	Cost per unit time and unit area exclusive of the BBT (circuity and terminal) costs.

References

Anily, S. and Federgruen, A. (1993) "Two-echelon distribution systems with vehicle routing costs and central inventories", *Opns. Res.* 41(1), 37-47.

Arrow, K. J., Karlin, S. and Scarf, H. (1958), *Studies in the Mathematical Theory of Inventory and Production*, Stanford Univ. Press, Stanford, CA.

Ball, M.O., Magnanti, T.L., Monma, C.L. and Nemhauser, G.L. (eds.) (1995), *Network Models*, Handbooks in Operations Research and Management Science, Vol 7, Elsevier, New York, N.Y.

Ball, M.O., Magnanti, T.L., Monma, C.L. and Nemhauser, G.L. (eds.) (1995a), *Network Routing*, Handbooks in Operations Research and Management Science, Vol 8, Elsevier, New York, N.Y.

Banks, J., Driscoll, W. and Stanford, R. (1982)" Design methodology for an airport limousine service" *Trans Sci.* 16, 127-148.

Bartholdi, J.J. III and Eisenstein, D.D. (1996) "A production line that balances itself", *Opns. Res.* 44(1), 21-34.

Beckmann, M. (1968) *Location Theory*, Random House, New York, N.Y.

Bertsimas, D.J., Jaillet, P. and Odoni, A.R. (1990) "A priori optimization", *Opns. Res.* 38(6), 1019-1033.

Bertsimas, D.J. and Van Ryzin, G. (1991) "A stochastic and dynamic vehicle routing problem in the Euclidean plane", *Opns. Res.* 39(4), 601-615.

Bhaskaran, S. and Daganzo, C.F. (1987) "Transportation and inventory considerations in single facility location", General Motors Research Laboratories Publication, Warren, Mich. (presented at ORSA/TIMS conference New Orleans, La, 1987).

Bhaskaran, S. and Kromer, R. (1986) "Internal memorandum on facilities location", General Motors Research Laboratories, Warren, Mich.

Blumenfeld, D.E., and Beckmann M. (1984) "Use of continuous space modeling to estimate freight distribution costs", General Motors Research Laboratories Publication GMR-4780, Warren, MI. Reprinted in *Trans. Res.* 19A(2) 173-187, 1985.

Blumenfeld, D.E., Hall, R.W. and Jordan, W.C. (1985) "Trade-offs between freight expediting and safety stock inventory costs", *J. Bus. Log.* 6, 79-100.

Blumenfeld, D.E., Burns, L.D., Diltz, J.D. and Daganzo, C.F. (1985a) "Analyzing trade-offs between transportation, inventory and production costs on freight networks", *Trans. Res. B* 19B(5), 361-380.

Blumenfeld, D.E., Burns, L.D. and Daganzo, C.F. (1986) "Synchronizing production and transportation schedules", General Motors Research Laboratories Publication GMR-5519, Warren, MI. Reprinted in *Trans. Res.* 25B(1) 23-27, 1991.

Blumenfeld, D.E., Burns, L.D., Frick, M., Daganzo, C.F. and Hall, R.W. (1987) "Reducing logistics costs at General Motors", *Interfaces* 17(1), 26-47.

Brandeau, M.L. and Chiu, S.S. (1989) "An overview of representative problems in location research", *Man. Sci.*, 35(6), 645-674.

Brown, G.R. and Graves, G.W. (1980), "Real time dispatch of petroleum tank trucks", Working paper 306, Western Management Science Institue, UCLA, Los Angeles, CA.

Buckingham, E. (1914) "On physically similar systems: Illustrations on the use of dimensional equations," *Physics Review* 4, 345-376.

Burns, L.D. (1986) Unpublished memoranda, General Motors Research Laboratories, Warren, Michigan.

Burns, L.D., and Daganzo, C.F. (1987) "Assembly line job sequencing principles", *Int. J. Prod. Res.* 25(1), 71-79.

Burns, L.D., Hall, R.W., Blumenfeld, D.E., and Daganzo, C.F. (1985) "Distribution strategies that minimize transportation and inventory costs", *Opns. Res.* 33(3), 469-490.

Campbell, J. (1990) "Location-allocation for distribution to a continuous demand with transshipments" School of Business Administration, U. of Missouri, St. Louis, MO. Reprinted in *Nav. Res. Log.* 39(5), 635-649, 1992.

Campbell, J. (1990a) "Locating transportation terminals to serve an expanding demand", *Trans. Res. B* 24B(3), 173-193.

Campbell, J. F. (1990b) "Freight consolidation and routing with transportation economies of scale", *Trans. Res.* 24B(5), 345-361.

Campbell, J. F. (1990c) " Designing logistics systems by analyzing transportation, inventory and terminal cost tradeoffs", *J. of Bus. Logistics* 11(1), 159-179,

Campbell, J. F. (1992) "Location-allocation for distribution systems with transshipments and transportation economies of scale", *Ann. Opns. Res.* 40, 77-79.

Campbell, J. F. (1993) "One-to-many distribution with transshipments: an analytic model", *Trans. Sci.* 27(4), 330-340.

Campbell, J. F. (1993a) "Continuous and discrete demand hub location problems", *Trans. Res.* 27B(6), 473-482.

Casco, D.O., Golden, B.L. and Wasil, E.A. (1988) "Vehicle routing with backhauls: Models algorithms and case studies", in *Vehicle Routing: Methods and Studies* (B. L. Golden and A. A. Assad, editors), Elsevier Science, Amsterdam, The Netherlands.

Clarens, G. and Hurdle, V.F. (1975) "An operating strategy for a commuter bus system", *Trans. Sci.* 9, 1-20.

Daganzo, C.F. (1984a) "The length of tours in zones of different shapes", *Trans. Res. B* 18B, 135-146.

Daganzo, C.F. (1984b) "The distance traveled to visit N points with a maximum of C stops per vehicle: An analytic model and an application", *Trans. Sci.* 18(4), 331-350.

Daganzo, C.F. (1985) "Mathematical specification of transportation models", in *Measuring the Unmeasurable*, P. Nijkamp, M.Leitner, and E. Wrigley eds., pp.663-678, NATO ASI Series D #22, Martinus Nijhoff, Dordrecht, The Netherlands.

Daganzo, C.F. (1985a) "Supplying a single location from heterogeneous sources", *Trans. Res. B* 19B(5), 409-420.

Daganzo, C.F. (1987) "Increasing model precision can reduce accuracy" *Trans. Sci.* 21(2), 100-105.

Daganzo, C.F. (1987a) "Modeling distribution problems with time windows", *Trans. Sci.* 21(3), 171-179.

Daganzo, C.F. (1987b) "Modeling distribution problems with time windows. Part II: Two customer types", *Trans. Sci.* 21(3), 180-187.

Daganzo, C.F. (1987c) "The break-bulk-role of terminals in many-to-many logistics networks", *Opns. Res.* 35(4), 543-555.

Daganzo, C.F. (1988) "A comparison of in-vehicle and out-of-vehicle freight consolidation strategies", *Trans Res. B* 22B(3), 173-180.

Daganzo, C.F. (1988a) "Shipment composition enhancement at a consolidation center", *Trans Res. B* 22B(2), 103-124.

Daganzo, C.F. (1990) "On the coordination of inbound and outbound schedules at a transportation terminal" *Proc. 11th Int. Symp. Trans. and Traffic Theory*, (M. Koshi, editor), pp. 379-390, Yokohama, Japan.

Daganzo, C.F. (2002) *A theory of supply chains*. Springer, Heidelberg, Germany.

Daganzo, C.F. (2004) "On the stability of supply chains", *Operations Research* 52(6), 909-921.

Daganzo, C.F. and Erera, A.L. (1999) "On planning and design of logistics systems for uncertain environments" in *New Trends in Distribution Logistics* (M.G. Speranza and P. Stahly, editors) Lecture Notes in Economics and Mathematical Systems, vol 480, pp. 3-21, Springer-Verlag, New York (1999).

Daganzo, C.F. and Hall, R.W. (1990) "A routing model for pickups and deliveries: No capacity restrictions on the secondary items", Institute of Transportation Studies Research Report UCB-ITS-RR-90-3, Univ. of California, Berkeley, CA. Reprinted in *Trans. Sci.* 27(4), 315-329, 1993.

Daganzo, C.F., Hendrickson, C.T. and Wilson, N.H.M. (1977) "An approximate analytic model of many-to-one demand responsive transportation", *Proc. 7th Int. Symp. on the Theory of Traffic Flow and Transportation,* (T. Sasaki and T. Yamaoka, editors), pp. 743-772, Kyoto, Japan.

Daganzo, C.F. and Newell, G.F. (1985) "Physical distribution from a warehouse: vehicle coverage and inventory levels", *Trans. Res. B* 19B(5), 397-408.

Daganzo, C.F. and Newell, G.F. (1986) "Configuration of physical distribution networks", *Networks* 16(2), 113-132.

Daganzo, C.F. and Newell, G.F. (1987) "Handling operations and the lot size trade-off", Institute of Transportation Studies Working Paper UCB-ITS-WP-87-9, Univ. of California, Berkeley, CA. Reprinted in *Trans. Res.* 27B(3) 167-184, 1993.

Daganzo, C.F. and Smilowitz, K. R. (2004) "Bounds and approximations for the transportation problem of linear programming and other scalable network problems" *Transportation Science* 38(3), 343-356..

Daskin, M.S. (1985) "Logistics: An overview of the state of the art and perspectives of future research", *Trans. Res. A*, 19A (5/6), 383-398.

Daskin, M.S. (1995) *Network and Discrete Location: Models, Algorithms and Applications*, Wiley, New York, USA.

Dejax, P.J. and Crainic, T.G. (1987) "A review of empty flows and fleet management models in freight transportation", *Trans. Sci.* 21, 227-247.

Drezner, Z. and Hamacher, H.W. (2002) *Facility Location: Applications and Theory.* Springer, Berlin, Germany.

Dror, M., Ball, M. and Golden B. (1985) "A computational comparison of algorithms for the inventory routing problem", Working paper 166, Dept. of Management Sciences Univ. of Waterloo, Waterloo, Ontario, Canada.

Du, Y. (1993) "Fleet sizing and empty equipment redistribution for transportation networks", PhD thesis, Dept. of Ind. Eng. and Opns. Res., Univ. of California, Berkeley, CA.

Du, Q., Faber, V. and Gunzburger, M. (1999) "Centroidal Voronoi tessellations: applications and algorithms", *SIAM Review*, 41(4): 637-676.

Eilon, S., Watson-Gandy, C.D.T. and Christofides, N. (1971) *Distribution Management: Mathematical Modelling and Practical Analysis*, Hafner, New York, N.Y.

Erera, A. L. (2000) "Design of large-scale logistics systems for uncertain environments" Ph.D. Thesis, Dept. of Industrial Engineering and Operations Research, U. California, Berkeley, CA.

Erlenkotter, D. (1988) "The general optimal market area model", Working paper 348, Western Management Science Institute, UCLA, Los Angeles, CA. Reprinted in *Ann. Opns. Res.* 18, 45-70, 1989.

Erlenkotter, D. (1990) "Ford Whitman Harris and the Economic Order Quantity Model", *Opns. Res.* 38(6), 937-946.

Fruin, J.J. (1971) *Pedestrian Planning and Design,* M.A.U.D.E.P., New York, N.Y.

Gallego, G., and Simchi-Levy, D. (1988) "On the effectiveness of direct shipping strategy for one warehouse multi-retailer R-systems", Dept. of Industrial Engineering Report, Columbia University, New York, N.Y.

Gendreau, M., Laporte, G. and Seguin, R. (1996) "Stochastic vehicle routing". *Euro. J. Opnl. Res.,* 88(1), 3-12.

Gleick, J. (1988) *Chaos: Making a New Science*, Penguin Books, New York, N.Y.

Glover, F. (1989) "Tabu search, part I", *ORSA J. on Computing* 1(3), 190-206.

Glover, F. (1990) "Tabu search, part II", *ORSA J. on Computing* 2(1). 4–32.

Golden B.L. and Yee J.R. (1979) "A framework for probabilistic routing", *AIIE Trans.* 11, 109-112.

Hall, R.W. (1984) "Travel distance through transportation terminals on a rectangular grid", *J. Opnl. Res. Soc.* 35, 1067-1078.

Hall, R.W. (1985) "Determining vehicle dispatch frequency when shipping frequency differs across suppliers", *Trans. Res. B* 19B(5), 421-432.

Hall, R.W. (1987) "Direct versus terminal freight routing on a network with concave costs", *Trans. Res. B* 21B(4), 287-298.

Hall, R.W. (1989) "Vehicle packing", *Trans. Res. B* 23B(2), 103-121.

Hall, R.W. (1989a) "Graphical Interpretation of the transportation problem", *Trans. Sci.* 23(1), 37-45.

Hall, R.W. (1989b) "Configuration of an overnight package air network", *Trans. Res. A* 23A(2), 139-150.

Hall, R.W. (1990) "Characteristics of multistop/multilevel vehicle routes", Institute of Transportation Studies Report, Univ. of California, Berkeley, CA. Reprinted in *Trans. Res.* 25B(6), 391-403, 1991.

Hall, R. W. (1992) "Pickup and delivery systems for overnight carriers", Transportation Center Working Paper 106, Univ. of California, Berkeley, CA. To appear in *Trans. Res.*

Hall, R.W. (1993) "Properties of vehicle routes with variable shipment sizes in euclidean plane", *Trans. Res. Rec.* 1413, 122-129.

Hall, R.W. (1993a) "Distance approximations for routing manual pickers in a warehouse", *IIE Transactions* 25(4), 76-87.

Hall, R. W. (1993b) "Design for local area freight networks", *Trans. Res.* 27B(2), 79-95.

Hall, R.W. and Daganzo, C.F. (1984) "Travel distance through transportation terminals on a grid: Alternative routing strategies", General Motors Research Publication GMR-4719. Warren, Mich.

Hall, R.W. and Daganzo, C.F. (1985) "Vehicle miles for a freight carrier with two capacity constraints", *Trans. Res. Rec.* 1038, 34-40.

Hall, R.W., Du, Y. and Lin, J. (1994) "Use of continuous approximations within discrete algorithms for routing vehicles: experimental results and interpretation", *Networks* 24(1), 43-56.

Hall, R.W. and Racer, M. (1995) "Transportation with common carrier and private fleets: system assignment and shipment frequency optimization", *IIE Transactions* 27, 217-225.

Han, A. (1984) "One-to-many distribution of nonstorable items: Approximate analytic models", PhD thesis, Dept of Civil Eng., Univ. of California, Berkeley, CA.

Han, A. and Daganzo, C.F. (1986) "Distributing nonstorable items without transshipments", *Trans. Res. Rec.* 1061, 32-41.

Han, A. and Daganzo, C.F. (1988) "Distributing nonstorable items: transshipments allowed", Paper presented at EURO IX/TIMS XXVIII Joint International Conference, Paris, France.

Harris, F.W. (1913) "How many parts to make at once", *Factory, The Magazine of Management* 10, 135-136,152.

Harris, F.W. (1913a) "How much stock to keep on hand", *Factory, The Magazine of Management* 10, 240-241, 281-284.

Hendrickson, C.T. (1978) "Approximate analytic performance of integrated transit components", PhD thesis, M.I.T., Cambridge, MA.

Hopfield, J.J. and Tank, D.W. (1985) "'Neural' computation of decisions in optimization problems", *Biol. Cybern.* 52, 141-152.

Horowitz, A.D. and Daganzo, C.F. (1986) "A graphical method for optimizing a continuous review inventory system", *Prod. Inv. Man.* 27 (4), 30-45.

Huang, M.D., Romeo, F., and Sangiovanni-Vincentelli, A. (1986) "An efficient general cooling schedule for simulated annealing", IEEE International Conference on Computer-Aided Design, Santa Clara, California.

Hurdle, V. O. (1973) "Minimum cost locations for parallel transit lines", *Trans. Sci.* 7, 340-350.

Hurdle, V.F. (1973a) "Minimum cost schedules for a public transportation route. I. Theory", *Trans. Sci.* 7(2), 109-137.

Hurdle, V.F. (1973b) "Minimum cost schedules for a public transportation route. II. Examples", *Trans. Sci.* 7(2), 138-157.

Jeng, C.Y. (1987) "Routing strategies for an idealized airline network", PhD thesis, Univ. of California, Berkeley, CA.

Jordan, W.C. (1982) "The impact of uncertain demand and supply on empty railroad car distribution", PhD thesis, Cornell University, Ithaca, N.Y.

Jordan, W.C. and Burns, L.D. (1984) "Truck backhauling on two terminal networks" *Trans. Res. B,* 18B(6), 487-503.

Karp, R.M. (1977) "Probabilistic analysis of partitioning algorithms for the traveling salesman problem", *Math. Opns. Res.* 2, 209-224.

Kiesling, M.K. (1995) "A comparison of freight distribution costs for combination and dedicated carriers in the air express industry", PhD thesis, Dept. of Civil Engineering, Univ. of California, Berkeley.

Klincewicz, J.G., Luss, H. and Pilcher, M. G. (1990) "Fleet size planning when outside carrier services are available", *Trans. Sci.* 24(3), 169-82.

Kirkpatrick, S., Gelatt, C.D., and Vecchi, M.P. (1983) "Optimization by simulated annealing," Sci. 220, 671-680.

Kunder, R. and Gudehus, T. (1975) "Mittlere wegzeiten beim ein-dimensionalen kommissionieren", *Zeitschrift fur Operations Research* 19, B53-B72.

Laarhoven, P.J.M. and Aarts, E.H.L. (1987) *Simulated Annealing: Theory and Applications,* D. Reidel Publishing Co., Dordrecht, The Netherlands.

Langevin, A., Mbaraga, P. and Campbell, J.F. (1995) "Continuous approximation models in freight distribution: an overview", Centre de Recherche sur les Transports Publication CRT-992, Universite de Montreal, Montreal, Canada. (*Trans. Res.* in press).

Langevin, A. and Saint-Mleux, Y. (1992) "A decision support system for physical distribution planning*",* J. *Decision Systems* 1, (2-3) 273-256.

Langevin, A. and Soumis, F. (1989) "Design of multiple-vehicle delivery tours satisfying time constraints", *Trans. Res. B* 23B(2), 123-138.

Larson, R.C. and Odoni, A.R. (1981) *Urban Operations Research*, Prentice-Hall, Englewood Cliffs, N.J.

Lin, S. (1965) "Computer solutions of the traveling salesman problem", *The Bell System Tech. J.*, 44, 2245-2267.

Lösch, A. (1954) *The Economics of Location* (translation of "Die Räumliche Ordnung der Wirtschaft, 2nd ed, 1944), Yale Univ. Press, New Haven, Conn.

Love, R.F., Morris, J.G. and Wesolowsky, G.O. (1988) *Facilities Location: Models and Methods,* North Holland, New York, N.Y.

Martin, J.D. (1989) "Third-party logistics comes of age", *Container News* pp.23-25, December, 1989.

Mitra D., Romeo F., and Sangiovanni-Vincentelli, A. (1986) "Convergence and finite-time behavior of simulated annealing", *Adv. in Appl. Probab.*, 18(3), 747-771.

Mitric, S. (1972) "Vertical transportation in tall buildings" Ph.D. thesis, Dept. of Civil Eng., Univ. of California, Berkeley, CA.

Newell, G.F. (1971) "Dispatching policies for a transportation route", *Trans. Sci.* 5, 91-105.

Newell, G.F. (1973) "Scheduling, location, transportation and continuum mechanics: some simple approximations to optimization problems", *SIAM J. Appl. Math.* 25(3), 346-360.

Newell, G.F. (1980) *Traffic Flow on Transportation networks*, MIT Press, Cambridge, Mass.

Newell, G.F. (1982) *Applications of Queueing Theory*, Second edition, Chapman Hall, London.

Newell, G.F. (1986) "Design of multiple vehicle delivery tours—III: Valuable goods", *Trans. Res. B* 20B(5), 377-390.

Newell, G.F. (1990) "Hiring temporary labor" (mimeo), Dep. of Civil Eng., Univ. of California, Berkeley, CA.

Newell, G.F. and Daganzo, C.F. (1986) "Design of multiple vehicle delivery tours—I: A ring-radial network", *Trans. Res. B* 20B(5), 345-364.

Newell, G.F. and Daganzo, C.F. (1986a) "Design of multiple vehicle delivery tours—II: Other metrics", *Trans. Res. B* 20B(5), 365-376.

Okabe, A., Boots, B. and Sugihara, K. (1992). *Spatial Tessellations: Concepts and Applications of Voronoi Diagrams.* Wiley, Chichester, UK

Ouyang, Y. and Daganzo, C.F. (2004) "Discretization and validation of the continuum approximation approach for terminal system design" Presented at the 2004 meeting of the *Trans. Res. Board.; Trans. Sci.* (in press).

Peterson, R. and Silver, E.A. (1979) *Decision Systems for Inventory Management and Production Planning,* Wiley, New York, N.Y.

Platzman, L.K. and Bartholdi, J.J., III (1989) "Spacefilling curves and the planar travelling salesman problem", *J. Assn. Comp. Mach.*, 36(4), 719-737.

Popken D. (1988) "Multiattribute, multicommodity flows in transportation networks" PhD thesis, Dept. of Industrial Engineering and Operations Research, U. California, Berkeley, CA.

Powell, W.B., Sheffi, Y. and Thiriez, S. (1984) "The dynamic vehicle allocation problem with uncertain demands", *Proc. 9th International Symposium on Transportation and Traffic Theory*, (J. Vollmuller and R. Hamerslag, editors), pp. 357-374, Delft, The Netherlands.

Press, W.H., Flannery, B.P., Teukolsky, S.A., and Vetterling, W.T., (1986) *Numerical Recipes: The Art of Scientific Computing*, Cambridge University Press, Cambridge, U.K.

Robuste, F., Daganzo, C.F. and Souleyrette, R. (1990), "Implementing vehicle routing models ", *Trans. Res. B* , 24(4), 263-286.

Schwarz, L.B. (editor) (1981) *Multi-level production/inventory control systems: theory and practice*, TIMS studies in the management sciences v 16, Elsevier North-Holland, Amsterdam, The Netherlands.

Sheffi, Y. (1985) *Urban Transportation Networks*, Prentice-Hall, Englewood Cliffs, N.J.

Smilowitz, K.R. (2001) "Design and operation of multimode, multiservice logistics systems," PhD thesis, Department of Civil and Environmental Engineering, U. California, Berkeley, CA.

Smilowitz, K.R. and Daganzo, C. F.(2004) "Cost modeling and design techniques for complex transportation systems" Working Paper 04-2004, IEMS Department, Northwestern University; (submitted for publication).

Steenbrink, P.A. (1974), *Optimization of Transport Networks*, John Wiley & Sons, New York, N.Y.

Taylor, J. R. (1997), *An Introduction to Error Analysis: The Study of Uncertainties in Physical Measurements* (2nd ed.), University Science Books, Sausalito, CA.

Vuchic, V. and Newell, G.F. (1968) "Rapid transit interstation spacing for minimum travel time", *Trans. Sci.* 2, 303-339.

Webb, I. (1989) "An integrated approach to strategic and tactical inventory/routing problems", PhD thesis, Massachusetts Institute of Technology, Cambridge, MA.

Weber, A. (1929) *On the Location of Industries* (translation of "Uber den Standort der Industrie", 1909), Univ. of Chicago Press, Chicago, Ill.

Welch, W.E. (1956) *Tested Scientific Inventory Control*, Management Publishing Co. Greenwich, Connecticut.

Wirasinghe, S.C. and Ghoneim, N.A. (1981) "Spacing of bus-stops for many to many travel demand", *Trans. Sci.* 15, 210-221.

Wirasinghe, C.S., Hurdle, V.F. and Newell, G.F. (1977) "Optimal parameters for a coordinated rail and bus transit system", *Trans. Sci.* 11(4), 359-74.

Zangwill, W.I. (1968) "Minimum concave cost flows in certain networks", *Man. Sci.* 14, pp. 429-450.

Zipkin, P.H. (2000) *Foundations of Inventory Management*, McGraw-Hill, New York, N.Y.

Appendix A: Some Properties of the TSP and the VRP

This appendix proves some properties of planar traveling salesman and vehicle routing problems. By this we mean, problems where the points x_i are randomly and uniformly located on a two-dimensional Cartesian region with distances given by a norm.

TSP

We look here for the expected distance of the optimum TS tour, d_{TSP}, in a region **R** of area $R = |\mathbf{R}|$ with N points on average. Point locations and tour lengths vary across realizations, e.g., from day to day. Locations are assumed to follow a homogeneous 2-dimensional point process with rate δ (points/area), such that $N = \delta R$. This process can take many forms; e.g., be a regular lattice,[1] a Poisson process, a clustered Poisson process or a hybrid of these. We shall show that $d_{TSP}/N \to k\delta^{-1/2}$ as $(N, R) \to \infty$ if both the process and the zone shape are fixed. The constant k depends on the norm and the process, but not on shape.

Assume for now that the zone shape, norm and process type are fixed, and that δ, R and N are parameters. Then, we know from dimensional analysis that the average distance per point, d_{TSP}/N, must be of the form, $\delta^{-1/2}f(N)$, where "f" is a function to be determined; see Buckingham (1914). We expect this function to be monotonic, and hence to have a limit, k, if it is bounded. In fact, the following is true.

Proposition 1: For square regions, the function "f" is bounded with $k > 0$.
Proof: Consider a square **R** composed of m^2 smaller squares of area S, and form a sub-optimal TS tour by forming optimal tours in the individual squares and linking them in sequence. A sequence of contiguous squares (sharing a side) exists if m is even. The expected length of such a sub-optimal tour across all possible realizations is bounded from above by the sum of the expected lengths of m^2 individual optimal tours, $m^2 N\delta^{-1/2}f(N) = m^2(NS)^{1/2}f(N)$, plus the expected length of the linkages. Since these links

---

[1] For homogeneous lattices the relation $N = \delta R$ is only meaningful for large regions, such that displacements of the lattice do not affect significantly the number of points in **R**. This restriction is not a problem for us, since the results in this appendix pertain to situations where $(N, R) \to \infty$.

join contiguous squares, their combined length cannot exceed $\kappa m^2 S^{1/2}$, where κ is a norm-specific constant. Thus, an upper bound to the expected length of the complete tour is: $[m^2 N^{1/2} f(N) + \kappa m^2] S^{1/2}$. This, of course, also bounds from above the expected length of the optimum tour, which is: $(Nm^2)\delta^{-1/2} f(Nm^2) = m^2 N^{1/2} f(Nm^2) S^{1/2}$. Thus, $m^2 N^{1/2} f(N) + \kappa m^2 \geq m^2 N^{1/2} f(Nm^2)$ for any even m; or, equivalently, $f(N) + \kappa N^{-1/2} \geq f(Nm^2)$. We now see by fixing N and letting m go to infinity that "f" is bounded–i.e., with a finite limit. We also see that this limit must be positive since it must exceed the expected distance between nearest neighbors, which is itself positive. $\square$

Corollary: The limit $f(N) \to k$, exists and is the same for zones of all shapes.
Proof: We have established that for large N the average distance per point in a square tends to $k\delta^{-1/2}$ as $N \to \infty$. Consider now a fixed region **R** partitioned into $M = N^{1/2}$ elementary squares with $N^{1/2}$ points each on average, where M, $N \to \infty$. (Zones of any shape can be approximated in this way.) Now, form a tour by linking optimal elementary tours as in the previous proof. For 2-dimensional regions **R**, the length of the linkages is of order $O(MS^{1/2}) = O(M^{1/2} R^{1/2}) = O(N^{1/4} R^{1/2}) = O(N^{3/4})\delta^{-1/2}$. The expected distance per point for the resulting tour is therefore: $k\delta^{-1/2} + O(N^{3/4})\delta^{-1/2}/N$, where k is the limiting constant for a square. Since the second term in this expression is of a lesser order it can be neglected for $N \to \infty$. This proves the corollary. $\square$

An upper bound to k can be found by finding the expected length of heuristic tour construction methods. We describe here the swath heuristic in Daganzo (1984a). Let **R** be a square, with distances given with by an L_1 norm oriented with the sides, and points by a homogeneous Poisson process with density δ. Cover the square with a swath of width w, as shown in Fig. A1. The width should be an even integer sub-multiple of $R^{1/2}$. Visit the points in order of appearance by traveling along the swath, only deviating laterally to visit them; see figure. If w is small so we can ignore corner effects, the distance of a tour can be decomposed into a longitudinal distance (the length of the swath) and a transverse distance (the sum of all the lateral deviations). The longitudinal distance is R/w. The average lateral deviation is the average distance between two random points in a segment of length w; i.e., w/3. Thus, the expected transverse distance is $(\delta R)w/3$, the expected total tour distance, $R/w + \delta R w/3$, and the expected distance per point: $1/(\delta w) + w/3$. This expression is minimized by $w = w^* \equiv (3/\delta)^{1/2}$, which yields for the average distance per point: $(4/3)^{1/2}\delta^{-1/2} \cong 1.15\ \delta^{-1/2}$.

Note that w* becomes small compared with the dimensions of **R** as δ increases. Thus, for δ → ∞ we can carve **R** with a swath of length arbitrarily close to w*, and we can conclude that k < 1.15 for the L_1 norm. For the Euclidean metric, similar derivations yield k < 0.87.

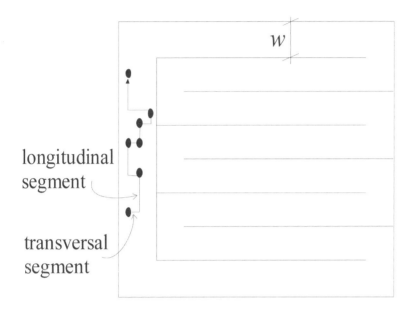

Fig. A1   The swath heuristic for the TSP

The formula also applies to rectangles and other geometric shapes that can be carved into a swath of near-optimum width even if δ is not large. But the formula does not apply to very narrow rectangles of length L and width *l*, if their width is considerably smaller than $2w^* = 2(3/\delta)^{1/2}$; i.e., if $\delta l^2 \ll 12$. In this case, which is of importance for the VRP, we can instead choose w = *l*/2 and construct a tour by traveling up and back the length of the rectangle just once. The expected distance for such a tour is 2L + N*l*/6, so that the average distance per point is 2L/N + *l*/6 = 2L/δR + *l*/6 = 2/δ*l* + *l*/6. Thus, a general TSP distance approximation formula for rectangles is:

$$d_{TSP}/N \cong 2/\delta l + l/6, \quad \text{if } \delta l^2 < 12, \text{ and}$$
$$\cong 1.15\, \delta^{-1/2}, \quad \text{otherwise.}$$

The top part of this formula is more accurate than the bottom, and its precision improves as *l* is reduced. (This should be intuitive: if *l* → 0 with fixed L and δ, the formula predicts d_{TSP} = 2L, which is the exact result.)

VRP

A VRP requires all the demand information for a TSP (points, metric, etc...) plus some information about the vehicle fleet. We shall only use here the vehicle capacity V (items/vehicle) and the depot location. Additional information such as vehicle speed and stopping time is relevant for some applications. We also need information about the items demanded at each point. We shall assume that customer demands are independent across customers, and given by a distribution that can be characterized by its mean, $v \ll V$.

If we assume momentarily that **R** is a circle and look for the expected optimum VRP distance d_{VRP} under the Euclidean norm, the problem can be characterized with only 5 parameters: R, N, v, V and the distance from the depot to the center of the circle, r. Therefore, to be dimensionally correct, its solution must be of the form: $(d_{VRP}/R^{1/2}) = f(r/R^{1/2}, N, v/V)$. We shall now show that when $N \to \infty$ with all else constant solutions to the VRP can be constructed with expected distance given by $2rN(v/V) + k_{VRP}(RN)^{1/2}$, and that this expression is near-optimal; i.e.,

$$d_{VRP}/N \cong 2r(v/V) + k_{VRP}\delta^{-1/2} \qquad \text{for } \delta \to \infty.$$

The first term on the right side is a "line-haul distance" needed to reach the customers from the depot, which depends on r, and the second term a "local distance" for detours, which depends on the characteristic distance between points $\delta^{-1/2}$. This expression holds for non-circular regions if r is interpreted as the distance from an item's demand point to the depot, averaged across all items. To establish these facts we first introduce some upper and lower bounds.

Proposition 2: An upper bound to the VRP distance is $2rN(v/V) + k_{TSP}(RN)^{1/2}$.
Proof: Consider the following TSP partitioning strategy. For each realization of a problem construct an optimum TST and number all the items in the order they would be collected, from 1 to S (the number of items for the realization). Send vehicle 1 to collect items 1 to V (in order), vehicle 2 to collect items V+1 to 2V, vehicle 3 to collect items 2V+1 to 3V, etc. until all items have been collected. The length of the resulting VRP routes is obviously bounded from above by the sum of d_{TSP} and the combined length of all the "line-haul" connectors between the TST and the depot. Consider now a similarly constructed VRP solution that starts with item 2 and ends with item 1. The new distance too, will be bounded from above by the sum of d_{TSP} and the combined length of the new connectors. If we construct S

VRP solutions in this manner we know that their average distance d_{avg} must bound from above the smallest distance in the collection; and hence the optimum VRP distance. Note that the combined length of all the connectors is the sum of the round trip distances from all items to the depot. Thus, if we now average across all realizations both d_{avg} and the optimum VRP distances, the former continues to bound the latter. Furthermore, the average length of the connectors of an average vehicle tour is $2r(Nv/V)$, as per the definition of r. Thus, the average of d_{avg} is $2r(Nv/V) + k_{TSP}(RN)^{1/2}$. □

Proposition 3: The two components of the upper bound, $2r(Nv/V)$ and $k_{TSP}(RN)^{1/2}$, are lower bounds to d_{VRP}.
Proof: The TSP portion, $k_{TSP}(RN)^{1/2}$, is a lower bound because the TSP solution solves optimally a less constrained VRP problem, with $V = \infty$.

To see that the line-haul portion is a lower bound, consider an optimum VRP solution and V identical copies of one of its individual vehicle tours. For each copy measure the two connectors between the depot and one of the items. In all cases the length of the tour is bounded from below by the length of these connectors. Therefore, it is bounded from below by their average. Since this is true for all the tours in the VRP solution, we see that the average vehicle tour length is bounded from below by 2r, and that $2r(Nv/V)$ must be a lower bound to d_{VRP}. □

The line-haul lower bound is tighter than the local component if $2r(Nv/V) > k_{TSP}(RN)^{1/2}$; i.e., if $2r/(V/v) > k_{TSP}(R/N)^{1/2} = k_{TSP} \, \delta^{-1/2}$. This inequality holds if $\delta > [k_{TSP}(V/v)/(2r)]^2$. When δ is small the reverse is true. We now show that tours with less local distance than predicted by the TSP partitioning method can be constructed, if δ is sufficiently large.

Consider a region **S** with an external depot, as shown in Fig. A2, where distances are given by a ring-radial metric centered at the depot. The region is first partitioned with a set of concentric rings of width L. A sweep is then performed on each ring to define the delivery zones. A zone is defined each time the item count reaches V. These zones are then divided longitudinally in two equal strips of length L and width w. Each vehicle is then routed from the depot to the nearest customer in its zone (in the radial direction) and then visits all the customers in the current strip in order of increasing radial distance. On reaching the end, the vehicle swaps strips by moving transversely along the ring and then visits the remaining customers in reverse order.

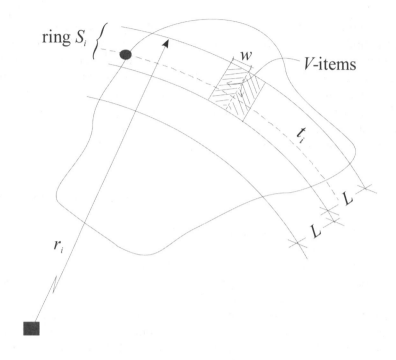

Fig. A2   The ring and sweep strategy for the VRP

The simplest result arises when the depot is so far from the region that distances inside each zone are given by an L_1 norm. In this case the VRP distance can be neatly decomposed into radial and transverse distances, and this is exploited below. Let N_i and D_i be the number of customers and the total demand in ring i, and R_i and T_i the corresponding radial and transverse distances. We use angled brackets "$<>$" to denote the average of these quantities across all realizations. They satisfy:

$<N_i> = \delta R_i$ (where R_i is the area of S_i –the portion of S in ring i),

$<D_i> = v\delta R_i$,

$<R_i> \leq 2r_i(v/V)\delta R_i$ (where r_i is the distance from the outer edge of ring i to the depot),

$<T_i> \leq t_i\,(V/v)/6$ (where t_i is the length of the longest arc in S_i).

The two inequalities are also good approximations. The second inequality requires some explanation. It assumes that S is large compared with a typical delivery zone. In this case most zones are away from the boundary and rectangular, as the one shown in Fig. A2, and a lateral move requires on average 1/6 of the width of the zone. Since the average number of zones in a ring is $\langle \lceil D_i / V \rceil \rangle \geq \langle D_i \rangle / V$ (we can ignore rounding effects for large S) the average distance for a move is approximately $(1/6)t_i V/\langle D_i \rangle$. We expect approximately $\langle D_i \rangle / v$ moves–one per customer visit. Therefore, the expected total lateral distance is $\langle T_i \rangle \cong t_i (V/v)/6$. We now show that this quantity is actually an upper bound.

Proof: If an individual tour has 1 or more stops on each strip, the number of lateral moves is n-2, where n is the number of stops in the tour. One of these moves crosses both strips and is therefore equivalent to two single-strip moves. Thus, there are n-1 single moves, with an average total distance of $(n-1)l/6$, where l is the width of the zone. If an individual tour has fewer than 1 stop on either of the strips its transverse distance is less. Thus, $(n-1)l/6$ is an upper bound to the average transverse distance of any tour conditional on n and l. Since the random variables n and l are independent (delivery lot sizes, which influence n, are independent of location) an upper bound to the unconditional average transverse distance of a tour is $\langle n-1 \rangle \langle l \rangle / 6$. Furthermore, since one of the points in a tour may also be visited by the previous tour (if the load has to be split) we see that $\langle n-1 \rangle$ is bounded from above by the average number of points in a tour; i.e., the ratio the number of points in the ring $\langle D_i \rangle / v$ and the average number of tours in the ring, $\langle \#_i \rangle$. Recall that $\langle \#_i \rangle = \langle \lceil D_i / V \rceil \rangle \geq \langle D_i \rangle / V$. Clearly, an upper bound to the average transverse distance of a tour is: $(V/v)\langle l \rangle / 6$. Since the average total transverse distance in ring i, $\langle T_i \rangle$, is the product of the average distance per tour and $\langle \#_i \rangle$, it follows that $\langle T_i \rangle \leq \langle \#_i \rangle (V/v)\langle l \rangle / 6$. Note now from the definition of t_i that the average width of a zone $\langle l \rangle$ is bounded from above by $t_i /\langle \#_i \rangle$. Hence, $\langle T_i \rangle \leq t_i(V/v)/6$. $\square$

The expected total VRP distance is the sum of the distance for all rings. When L is sufficiently small compared with the diameter of S (we continue to ignore boundary effects) this can be expressed as:

$$\langle \text{Total VRP distance} \rangle \cong \sum 2r_i(v/V)\delta R_i + t_i(V/v)/6$$
$$\cong (2r+L)(v/V)\delta R + (R/L)(V/v)/6.$$

The first terms on both sides of the last equality are equal because the round trip distance to the edge of a ring is the round trip distance to its center plus the width of the ring. The second terms are equal because the area of S_i approximately equals the product of L and t_i.

If we now choose L to minimize the right side we find:

$$L^* = (V/v)(6\delta)^{-1/2}$$

and

Total VRP distance $\cong 2r(v/V)\delta R + 2R(\delta/6)^{1/2} \cong 2r(Nv/V) + 0.82(RN)^{1/2}.$

Since the expected number of tours is Nv/V, the expected distance per tour is $2r + k\delta^{-1/2}$ (V/v), where k = 0.82. If V >> v, the factor V/v is an approximation for the average number of stops in a tour. This is the result on which Eq. (4.1) of the text is based. The text assumed that the number of stops was fixed--as if V was a fixed integer multiple of v: V/v = C. The logic of this appendix also leads to Eq. (4.1) for other norms and inhomogeneous metrics. For the Euclidean metric the value is k $\cong$ 0.57 (Daganzo 1984b). This is considerably lower than the TSP partitioning value, k_{TSP}.[2]

---

[2] Current estimates of the Eucledian k_{TSP} based on computer simulations exceed 0.7. A lower bound to k_{TSP} based on the average distance from a random point to its nearest and second nearest neighbors is 0.625.

Appendix B

Simulated Annealing

Simulated Annealing (SA) belongs to a general class of "probabilistic hill climbing" algorithms whose goal is to approximate global optima in complex combinatorial optimization problems. Based on an analogy with annealing in solids (Kirkpatrick et al., 1983), SA can be described as a process which "... first 'melts' the system being optimized at a high effective 'temperature', then (slowly) lowers the 'temperature' in stages until the system 'freezes' and no further changes occur." Its objective is finding a configuration (or state) for which a certain cost function takes its minimum value, (i.e. a minimum energy state).

The SA algorithm generates new configurations with some probabilistic rules, and either accepts or rejects them depending on their relative cost. Unlike other iterative improvement algorithms, however, SA will sometimes accept configurations that increase cost if the result of a certain random acceptance rule is positive. The probability of acceptance is controlled by a parameter, T, analogous to the temperature in physics.

Given an initial configuration, s_1 , with cost $c(s_1)$, and initial temperature, $T=T_1$, the SA algorithm changes the configuration and the temperature with each iteration, t=1,2,..., in the following way (the terminology is adopted from Mitra et al., 1986): A *generate* function chooses a candidate configuration to jump from the current configuration, s_t , an *accept* function determines whether the candidate should be accepted, and an *update* function changes the temperature from T_t to T_{t+1} ; the latter function also updates the configuration.

If the algorithm is in a certain configuration at time t, the function *generate* randomly selects a new configuration from a set of feasible "neighbor" configurations. The design of this function must be customized for each particular problem – a function will be "theoretically sound" (i.e. guaranteed to converge when combined with other theoretically sound accept and update functions) if *it can reach any state from the initial state by jumping from neighbor to neighbor in a finite number of moves, and if the neighbors are chosen with equal probability.* This is a sufficient, although not necessary condition. Lin (1965) has proposed a set of moves for the TSP.

Although several accept functions are theoretically sound, the most common is that proposed by Kirkpatrick et al. (1983), which is based on

Boltzmann's probability density distribution (the result is known as the Metropolis algorithm). If the cost of the new configuration, $c(s_{t+1})$, is less than the cost of the old configuration, $c(s_t)$, the new configuration is accepted with probability one. Otherwise, it is accepted with probability:

$$\exp\left\{-\frac{[c(s_{t+1}) - c(s_t)]}{T_t}\right\}.$$

The time t is increased by one unit regardless.

The *update* function defines the "cooling schedule" which reduces the temperature from T_t to T_{t+1} . A schedule is theoretically sound if it decreases the temperature very slowly (i.e. no faster than $(\log(t))^{-1}$ for schedules in which the temperature is reduced at every iteration), and has a sufficiently large initial temperature.

With theoretically sound *generate*, *accept* and *update* functions, the simulated annealing algorithm has been shown to converge in probability to a minimum cost configuration (see Laarhoven and Aarts, 1987, and Mitra et al., 1986, for details). However, this results in prohibitively long execution times. Although guidelines exist to accelerate the cooling schedule in practice (see Huang et al., 1986, and Laarhoven et al., 1987, for example), execution times can still be long. For a specific problem type some experimentation is desirable to determine practical cooling schedules; i.e. settings for the initial temperature, the reduction factor, and a convergence criterion.

For effective application of simulated annealing there must be: 1) a quick way of calculating the cost of a configuration, and 2) a quick way of altering the system. For the classical TSP, the cost function is the tour length. The SA algorithm then tries variations on this tour and checks for improvements in tour length. The following implementation for the TSP (Press et al., 1986) was adopted for the experiments in Robuste et al., (1990):

1. Configuration. Points are numbered i=1,2...,N and each has coordinates (X_i, Y_i). A configuration is any permutation of the set {1,2,...,N}. Its cost is the sum of the distances between the N consecutive pairs of points,
2. Generating rearrangements (Lin, 1965). Either 1) a (random) section of the tour is removed and replaced with the same points running in the opposite order (a reversal), or 2) a (random) section of the tour is

removed and replaced between two points in another randomly chosen part of the tour (a transport),

3. Annealing schedule. Initially the temperature is set equal to a large initial value, T_1. The temperature is maintained constant for several trials (i.e. for a "temperature step"), and then is geometrically reduced by a constant factor, α .

For this annealing schedule to guarantee convergence in probability, the number of iterations in a temperature step must be large enough for the underlying Markov Chain to have reached a steady state. While methods exist to estimate whether this steady state has been reached, they are likely to result in long temperature steps – and execution times – especially so after the temperatures have been lowered.

Index

E

F

G

T

Printing and Binding: Strauss GmbH, Mörlenbach